T0155629

# NEUROMETHODS

**Series Editor**
**Wolfgang Walz**
**University of Saskatchewan,**
**Saskatoon, SK, Canada**

*Neuromethods* publishes cutting-edge methods and protocols in all areas of neuroscience as well as translational neurological and mental research. Each volume in the series offers tested laboratory protocols, step-by-step methods for reproducible lab experiments and addresses methodological controversies and pitfalls in order to aid neuroscientists in experimentation. *Neuromethods* focuses on traditional and emerging topics with wide-ranging implications to brain function, such as electrophysiology, neuroimaging, behavioral analysis, genomics, neurodegeneration, translational research and clinical trials. *Neuromethods* provides investigators and trainees with highly useful compendiums of key strategies and approaches for successful research in animal and human brain function including translational "bench to bedside" approaches to mental and neurological diseases.

# Cerebrovascular Reactivity

## Methodological Advances and Clinical Applications

Edited by

### Jean Chen

*Baycrest Centre for Geriatric Care, University of Toronto, Toronto, ON, Canada*

### Jorn Fierstra

*Department of Neurosurgery, University Hospital Zurich, Zürich, Zürich, Switzerland*

 Humana Press

*Editors*
Jean Chen
Baycrest Centre for Geriatric Care
University of Toronto
Toronto, ON, Canada

Jorn Fierstra
Department of Neurosurgery
University Hospital Zurich
Zürich, Zürich, Switzerland

ISSN 0893-2336          ISSN 1940-6045   (electronic)
Neuromethods
ISBN 978-1-0716-1765-6          ISBN 978-1-0716-1763-2   (eBook)
https://doi.org/10.1007/978-1-0716-1763-2

© Springer Science+Business Media, LLC, part of Springer Nature 2022, Corrected Publication 2022
This work is subject to copyright. All rights are reserved by the Publisher, whether the whole or part of the material is concerned, specifically the rights of translation, reprinting, reuse of illustrations, recitation, broadcasting, reproduction on microfilms or in any other physical way, and transmission or information storage and retrieval, electronic adaptation, computer software, or by similar or dissimilar methodology now known or hereafter developed.
The use of general descriptive names, registered names, trademarks, service marks, etc. in this publication does not imply, even in the absence of a specific statement, that such names are exempt from the relevant protective laws and regulations and therefore free for general use.
The publisher, the authors, and the editors are safe to assume that the advice and information in this book are believed to be true and accurate at the date of publication. Neither the publisher nor the authors or the editors give a warranty, expressed or implied, with respect to the material contained herein or for any errors or omissions that may have been made. The publisher remains neutral with regard to jurisdictional claims in published maps and institutional affiliations.

This Humana imprint is published by the registered company Springer Science+Business Media, LLC part of Springer Nature.
The registered company address is: 1 New York Plaza, New York, NY 10004, U.S.A.

# Preface to the Series

Experimental life sciences have two basic foundations: concepts and tools. The *Neuro-methods* series focuses on the tools and techniques unique to the investigation of the nervous system and excitable cells. It will not, however, shortchange the concept side of things as care has been taken to integrate these tools within the context of the concepts and questions under investigation. In this way, the series is unique in that it not only collects protocols but also includes theoretical background information and critiques which led to the methods and their development. Thus it gives the reader a better understanding of the origin of the techniques and their potential future development. The *Neuromethods* publishing program strikes a balance between recent and exciting developments like those concerning new animal models of disease, imaging, *in vivo* methods, and more established techniques, including, for example, immunocytochemistry and electrophysiological technologies. New trainees in neurosciences still need a sound footing in these older methods in order to apply a critical approach to their results.

Under the guidance of its founders, Alan Boulton and Glen Baker, the *Neuromethods* series has been a success since its first volume published through Humana Press in 1985. The series continues to flourish through many changes over the years. It is now published under the umbrella of Springer Protocols. While methods involving brain research have changed a lot since the series started, the publishing environment and technology have changed even more radically. Neuromethods has the distinct layout and style of the Springer Protocols program, designed specifically for readability and ease of reference in a laboratory setting.

The careful application of methods is potentially the most important step in the process of scientific inquiry. In the past, new methodologies led the way in developing new disciplines in the biological and medical sciences. For example, Physiology emerged out of Anatomy in the nineteenth century by harnessing new methods based on the newly discovered phenomenon of electricity. Nowadays, the relationships between disciplines and methods are more complex. Methods are now widely shared between disciplines and research areas. New developments in electronic publishing make it possible for scientists that encounter new methods to quickly find sources of information electronically. The design of individual volumes and chapters in this series takes this new access technology into account. Springer Protocols makes it possible to download single protocols separately. In addition, Springer makes its print-on-demand technology available globally. A print copy can therefore be acquired quickly and for a competitive price anywhere in the world.

*Saskatoon, SK, Canada*                                                                                     *Wolfgang Walz*

# Preface

Cerebrovascular reactivity (CVR) reflects changes in cerebral blood flow in response to a vasoactive stimulus. It has been used as a marker of cerebrovascular health and vascular reserve capacity. This volume is aimed at providing a comprehensive overview of the methodology, physiology as well as contemporary and novel applications of CVR measurements.

The chapters start with an introduction to the neurophysiology, neuroimaging, and clinical methods for CVR measurement and expand to include the use of CVR methods in the study of aging, cerebrovascular dysfunction, dementia, and brain tumors. The final chapters outline recommendations for measurement protocols and future applications in clinical translation.

This volume is designed to provide researchers in Neuroscience and Neurology alike with an up-to-date and comprehensive resource on the measurement, interpretation, and application of CVR measurement. This compilation is expected to foster the increasing research interest in the role of cerebrovascular dysfunction in various brain diseases.

*Toronto, ON, Canada*  
*Zürich, Zürich, Switzerland*

*Jean Chen*  
*Jorn Fierstra*

# Contents

Series Preface . . . . . . . . . . . . . . . . . . . . . . . . . . . . . . . . . . . . . . . . . . . . . . . . . . . . .    *v*

Preface . . . . . . . . . . . . . . . . . . . . . . . . . . . . . . . . . . . . . . . . . . . . . . . . . . . . . . . . . .    *vii*

Contributors . . . . . . . . . . . . . . . . . . . . . . . . . . . . . . . . . . . . . . . . . . . . . . . . . . . .    *xi*

1    The Physiological Basis of Cerebrovascular Reactivity Measurements . . . . . . . . . .   1
*Olivia Sobczyk, James Duffin, Joseph A. Fisher, and David J. Mikulis*

2    Experimental Protocols in CVR Mapping . . . . . . . . . . . . . . . . . . . . . . . . . . . . . . .   19
*Marat Slessarev*

3    Molecular Imaging of Cerebrovascular Reactivity . . . . . . . . . . . . . . . . . . . . . . . . .   33
*Audrey P. Fan and Oriol Puig Calvo*

4    Measurement of Cerebrovascular Reactivity Using Transcranial Doppler . . . . . . .   59
*Leodante da Costa and Martin Chapman*

5    The Role of Cerebrovascular Reactivity Mapping in Functional MRI:
Calibrated fMRI and Resting-State fMRI . . . . . . . . . . . . . . . . . . . . . . . . . . . . . . .   75
*J. Jean Chen and Claudine J. Gauthier*

6    Hemodynamic Evaluation of Paradoxical Blood Oxygenation
Level-Dependent Cerebrovascular Reactivity with Transcranial Doppler
and MR Perfusion in Patients with Symptomatic Cerebrovascular
Steno-occlusive Disease . . . . . . . . . . . . . . . . . . . . . . . . . . . . . . . . . . . . . . . . . . . .   89
*Christiaan Hendrik Bas van Niftrik, Martina Sebök, Giovanni Muscas,
Aimée Hiller, Matthias Halter, Susanne Wegener, Luca Regli,
and Jorn Fierstra*

7    Cerebrovascular Reactivity (CVR) in Aging, Cognitive Impairment,
and Dementia . . . . . . . . . . . . . . . . . . . . . . . . . . . . . . . . . . . . . . . . . . . . . . . . . . . .   103
*Hanzhang Lu, Binu P. Thomas, and Peiying Liu*

8    Magnetic Resonance Imaging Methods for Assessment
of Hemodynamic Reserve in Chronic Steno-occlusive Cerebrovascular
Disease . . . . . . . . . . . . . . . . . . . . . . . . . . . . . . . . . . . . . . . . . . . . . . . . . . . . . . . . .   119
*Keith R. Thulborn, Laura Stone McGuire, Fady T. Charbel,
and Sepideh Amin-Hanjani*

9    Breath-Hold Cerebrovascular Reactivity Mapping for Neurovascular
Uncoupling Assessment in Primary Gliomas . . . . . . . . . . . . . . . . . . . . . . . . . . . . .   167
*Domenico Zacà, Shruti Agarwal, and Jay J. Pillai*

10   Clinical Translation of Cerebrovascular Reactivity Mapping . . . . . . . . . . . . . . . . .   185
*Manus J. Donahue*

11   Recent Advances and Future Directions: Clinical Applications
     of Intraoperative BOLD-MRI CVR..................................... 207
     *Giovanni Muscas, Christiaan Hendrik Bas van Niftrik, Martina Sebök,*
     *Alessandro Della Puppa, Luca Regli, and Jorn Fierstra*

Correction to: Recent Advances and Future Directions: Clinical Applications
of Intraoperative BOLD-MRI CVR ......................................... C1

*Index* ................................................................ *217*

# Contributors

SHRUTI AGARWAL • *Division of Neuroradiology, The Russell H. Morgan Department of Radiology and Radiological Science, Johns Hopkins University School of Medicine, Baltimore, MD, USA*

SEPIDEH AMIN-HANJANI • *Department of Neurosurgery, University of Illinois at Chicago, Chicago, IL, USA*

MARTIN CHAPMAN • *Department of Critical Care Medicine, Sunnybrook Health Sciences Center, Toronto, ON, Canada; Department of Anesthesia, University of Toronto, Toronto, ON, Canada*

FADY T. CHARBEL • *Department of Neurosurgery, University of Illinois at Chicago, Chicago, IL, USA*

J. JEAN CHEN • *Rotman Research Institute, Baycrest Centre for Geriatric Care, Toronto, ON, Canada; Department of Medical Biophysics, University of Toronto, Toronto, ON, Canada*

LEODANTE DA COSTA • *Division of Neurosurgery, Sunnybrook Health Sciences Center, Toronto, ON, Canada; Department of Surgery, University of Toronto, Toronto, ON, Canada*

MANUS J. DONAHUE • *Department of Radiology and Radiological Sciences, Vanderbilt University Medical Center, Nashville, TN, USA; Department of Neurology, Vanderbilt University Medical Center, Nashville, TN, USA; Department of Psychiatry, Vanderbilt University Medical Center, Nashville, TN, USA*

JAMES DUFFIN • *Department of Anesthesiology, Toronto General Hospital, University of Toronto, Toronto, ON, Canada*

AUDREY P. FAN • *Department of Biomedical Engineering, University of California, Davis, Davis, CA, USA; Department of Neurology, University of California, Davis, Davis, CA, USA*

JORN FIERSTRA • *Department of Neurosurgery, University Hospital Zurich, University of Zurich, Zurich, Switzerland; Clinical Neuroscience Center, University Hospital Zurich, University of Zurich, Zurich, Switzerland*

JOSEPH A. FISHER • *Department of Anesthesiology, Toronto General Hospital, University of Toronto, Toronto, ON, Canada*

CLAUDINE J. GAUTHIER • *Department of Physics, Concordia University, Montreal, QC, Canada; Montreal Heart Institute, Montreal, QC, Canada*

MATTHIAS HALTER • *Department of Neurosurgery, University Hospital Zurich, University of Zurich, Zurich, Switzerland; Clinical Neuroscience Center, University Hospital Zurich, University of Zurich, Zurich, Switzerland*

AIMÉE HILLER • *Department of Neurosurgery, University Hospital Zurich, University of Zurich, Zurich, Switzerland; Clinical Neuroscience Center, University Hospital Zurich, University of Zurich, Zurich, Switzerland*

PEIYING LIU • *The Russell H. Morgan Department of Radiology & Radiological Science, Johns Hopkins University School of Medicine, Baltimore, MD, USA*

HANZHANG LU • *The Russell H. Morgan Department of Radiology & Radiological Science, Johns Hopkins University School of Medicine, Baltimore, MD, USA; F.M. Kirby Center for Functional Brain Imaging, Kennedy Krieger Institute, Baltimore, MD, USA;*

*Department of Biomedical Engineering, Johns Hopkins University School of Medicine, Baltimore, MD, USA*

LAURA STONE McGUIRE • *Department of Neurosurgery, University of Illinois at Chicago, Chicago, IL, USA*

DAVID J. MIKULIS • *Department of Medical Imaging, Toronto Western Hospital, University of Toronto, Toronto, ON, Canada*

GIOVANNI MUSCAS • *Department of Neurosurgery, University Hospital Zurich, University of Zurich, Zurich, Switzerland; Clinical Neuroscience Center, University Hospital Zurich, University of Zurich, Zurich, Switzerland; Department of Neurosurgery, Careggi University Hospital, University of Florence, Florence, Italy*

JAY J. PILLAI • *Division of Neuroradiology, The Russell H. Morgan Department of Radiology and Radiological Science, Johns Hopkins University School of Medicine, Baltimore, MD, USA; Department of Neurosurgery, Johns Hopkins University School of Medicine, Baltimore, MD, USA*

ORIOL PUIG CALVO • *Department of Clinical Physiology and Nuclear Medicine, Zeeland University Hospital, Koge, Denmark*

ALESSANDRO DELLA PUPPA • *Department of Neurosurgery, Careggi University Hospital, University of Florence, Florence, Italy*

LUCA REGLI • *Department of Neurosurgery, University Hospital Zurich, University of Zurich, Zurich, Switzerland; Clinical Neuroscience Center, University Hospital Zurich, University of Zurich, Zurich, Switzerland*

MARTINA SEBÖK • *Department of Neurosurgery, University Hospital Zurich, University of Zurich, Zurich, Switzerland; Clinical Neuroscience Center, University Hospital Zurich, University of Zurich, Zurich, Switzerland*

MARAT SLESSAREV • *Division of Critical Care Medicine, Department of Medicine, Western University, London, ON, Canada*

OLIVIA SOBCZYK • *Department of Anesthesiology, Toronto General Hospital, University of Toronto, Toronto, ON, Canada*

BINU P. THOMAS • *Advanced Imaging Research Center, University of Texas Southwestern Medical Center, Dallas, TX, USA*

KEITH R. THULBORN • *Center for Magnetic Resonance Research, University of Illinois at Chicago, Chicago, IL, USA*

CHRISTIAAN HENDRIK BAS VAN NIFTRIK • *Department of Neurosurgery, University Hospital Zurich, University of Zurich, Zurich, Switzerland; Clinical Neuroscience Center, University Hospital Zurich, University of Zurich, Zurich, Switzerland*

SUSANNE WEGENER • *Clinical Neuroscience Center, University Hospital Zurich, University of Zurich, Zurich, Switzerland; Department of Neurology, University Hospital Zurich, University of Zurich, Zurich, Switzerland*

DOMENICO ZACÀ • *Siemens Healthcare, srl, Milano, Italy*

# Chapter 1

# The Physiological Basis of Cerebrovascular Reactivity Measurements

## Olivia Sobczyk, James Duffin, Joseph A. Fisher, and David J. Mikulis

## Abstract

The brain requires a continuous supply of oxygen, which in normal environments is directly related to cerebral blood flow. Accordingly, a number of mechanisms are in place to maintain blood supply in the face of challenges such as variations in arterial blood pressure, hypoxemia, and vascular occlusions. These include the recruitment of collateral flow such as via the circle of Willis, the pial networks, and anastomoses between penetrating arterioles and capillaries. In addition, physiological mechanisms adjust the distribution of flow: (1) *autoregulation* to maintain flow during supply pressure changes, (2) *neurovascular coupling* to increase flow in regions of neuronal activity, and (3) *hypoxia-induced vasodilatation*. The common effector pathway for physiological mechanisms is the adjustment of vascular diameter. In the absence of steno-occlusive vascular disease, the flow response to a vasodilatory challenge, cerebrovascular reactivity (CVR), interrogates the physiological response. In the presence of steno-occlusive disease, CVR reflects both the health of the physiological regulators and the availability of collateral flow. Even in health, CVR varies between anatomical regions. As a result, CVR must be normalized for the region before it can be interpreted.

**Keywords** Cerebrovascular reactivity, Cerebral blood flow, Autoregulation, Physiology, Steno-occlusive disease, Cerebrovascular

## 1 Introduction

### 1.1 Supply of Cerebral Blood Flow

Brain tissue stores little energy resulting in a continuous demand for oxygen ($O_2$) and glucose, which, in turn, requires a continuous and well-regulated blood flow for normal brain function and cell viability. Indeed, cerebral oxygen consumption accounts for approximately 20% of the total resting body oxygen consumption and 15–20% of the total cardiac output [1, 2]. Cerebral blood flow (CBF) on average exceeds 50 mL/100 g tissue/min (approximately 70 mL/100 g/min in gray matter (GM) and 20 mL/100 g/min in white matter (WM) supporting an overall cerebral metabolic rate of oxygen ($CMRO_2$) of 3.5 mL/100 g tissue/min) [2]. At normal resting levels of CBF, the brain extracts approximately 40% of the $O_2$ and 10% of the glucose from the arterial

Jean Chen and Jorn Fierstra (eds.), *Cerebrovascular Reactivity: Methodological Advances and Clinical Applications*, Neuromethods, vol. 175, https://doi.org/10.1007/978-1-0716-1763-2_1, © Springer Science+Business Media, LLC, part of Springer Nature 2022

blood to satisfy its energy requirements. With blood flow reduced below a certain threshold, cellular function declines, and remaining energy is used to just maintain cellular integrity. If blood flow falls further or the duration is extended, the cells die from ischemia, hypoxia, or apoptosis [3].

In normal physiological conditions, the demand for CBF is met by changes in the vascular resistance in parenchymal arterioles, with little variation in overall flow. The large pial arteries on the surface of the cortex contain multiple layers of vascular smooth muscle cells. These vessels branch into penetrating arterioles containing a single layer of vascular smooth muscle cells [4] and enter the cortical parenchyma where capillary pericytes may control the flow [5, 6]. The tone of the smooth muscle is the final effector in (1) *autoregulation* which mitigates against variations in brain perfusion pressure [7, 8] and (2) *neurovascular coupling* which increases local blood flow in response to increased neuronal metabolic demand [6, 9, 10].

## 2    Anatomical Protective Features of Cerebral Circulation

Upon abrupt partial or complete occlusion of a systemic supply artery, blood flow can be rerouted around the circle of Willis to redistribute blood flow [11] and maintain downstream perfusion. In the longer term, the development of collateral pial and intracerebral vessels may augment the compensation [12, 13]. The effects of the variety of perfusion pressures and vascular resistances in a network of collateral vessels can be modeled as a simplified mechanism of perfusion pressure and resistance as shown in Fig. 1.

## 3    Physiological Compensation, Neovascularization, and Remodeling

Other physiological factors also affect CBF. These include acute changes in arterial blood gases [14, 15], such as hypoxia [16] and hypercapnia [17], as well as decreases in hemoglobin (anemia) [18, 19]. In the presence of hypoxia and anemia, vascular tone decreases to maintain adequate $O_2$ supply. Long-term changes in CBF occur in chronic anemia [20] including sickle cell anemia [21], altitude acclimatization [22], and chronic hypoxia [23]. These increases in flow required to maintain $O_2$ delivery ($DO_2$) are accompanied by a multitude of adaptive changes orchestrated via the HIF-1 alpha pathway [24]. Over the long term, the cerebral vasculature remodels to provide larger-diameter vessels and accommodate higher CBF. Interestingly, these changes reverse with correction of the anemia [25].

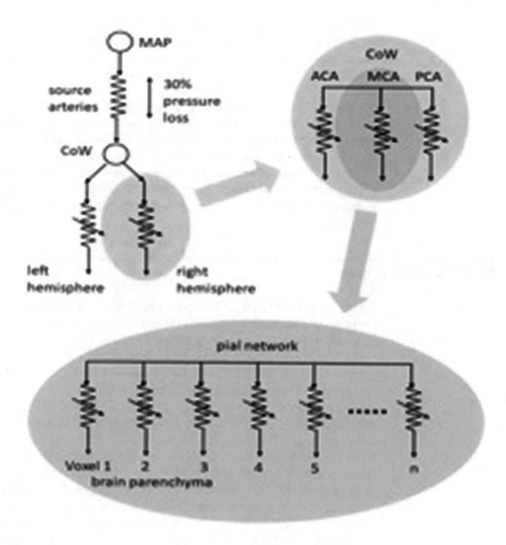

**Fig. 1** A conceptual schematic of cerebral blood flow pathways and resistances [57]. Resistances are represented with electrical resistance symbols with variability indicated by an arrow. Mean arterial pressure (MAP) forces flow through the high resistance of the source arteries to anastomose around the circle of Willis (CoW) where they are distributed to vascular territories, via the anterior, middle, and posterior cerebral arteries (ACA, MCA, and PCA, respectively) in each hemisphere. The interconnected pial networks supply the brain parenchyma via penetrating arterioles, with the perfusion of the deeper brain tissues reflecting the pattern of vasclar resistances and perfusion pressures

## 4  Cerebrovascular Pathology and CBF Overview

Obstruction of an artery, either by thrombus or embolus, is the commonest cause for local ischemic damage. Brain tissue deprived of blood supply undergoes necrosis or infarction (stroke) [26]. Slowly progressive steno-occlusive diseases may also simply produce a chronic hypoperfusion leading to thinning of the cortex in GM [27] and WM hyperintensities (WMH) [28, 29], both leading to neurological and cognitive decline [30].

Steno-occlusive disease pathologies vary from extracranial or intracranial focal stenoses to more progressive vasculopathies that affect multiple vessels. These obstructions co-exist with other developmental changes that tend to mitigate downstream ischemia such as selective, regional decreases of vascular resistance and the development of collateral circulation [12]. Clinical manifestation of steno-occlusive disease depends on the time course of the occlusion, its location, extent, and the availability of recruitable collateral circulation [12]. The latter, more than the degree of local vascular impairment, plays the crucial role in determining the risk of stroke [12, 31].

## 5   The Collateral Circulation

There are several levels of collateral circulation: the circle of Willis redistributes blood flow in the presence of an occlusion or stenosis in a major extracranial feeding vessel. Distal to the circle of Willis, there is potential cross perfusion between the intracranial vascular territories in the pial vessels at their borders over the convexity of the cortex. These secondary collateral systems may be inherent easily recruited structures in some people. In others, they may be enhanced—or even develop de novo—over time in response to tissue sensing of vascular inadequacy [32]. Arterioles branching from the pial network enter the cortex at right angles to the brain surface, mostly perfusing silos of tissue. However, with age and with reductions in perfusion pressures, anastomoses develop between vessels in these tissue silos, between adjacent silos and adjacent vascular territories. As such, collateral flow becomes possible within all levels of the vascular tree, pial arterioles and penetrating arterioles, down to capillaries as shown in Fig. 2 [6].

The availability of collateral circulation to supply compensatory blood flow depends on the site of occlusion, the genetics (the presence of in situ connections between vascular beds such as between vascular territories of the pial vessels), and the speed of occlusion (opportunity to develop collateral vessels). The effect of a sudden occlusion depends on the site of the occlusion and the availability of immediately recruitable intrinsic anastomoses. But if the occlusion develops more slowly, collateral vessels may develop by arteriogenesis. Animal experiments show that collateral compensation may evolve over time [12], and even so, its presence does not guarantee the adequacy of compensatory flow to meet tissue need when stressed.

In several cerebral diseases associated with inflammation and hemorrhage, the primary site of dysfunction is the neurovascular unit [33]. Other cerebrovascular pathological conditions, such as

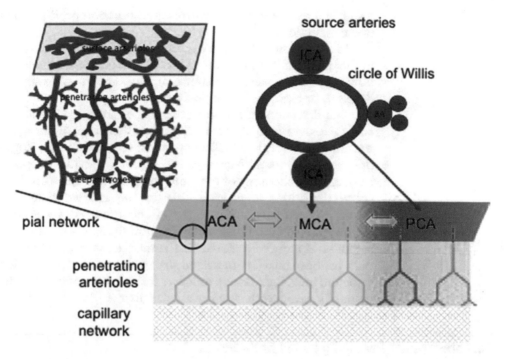

**Fig. 2** A conceptual schematic of the organization of cerebral blood flow (CBF) distribution and possible collateral blood flow paths. CBF is supplied by extracranial arteries: right and left internal carotid arteries (ICA) and right and left vertebral arteries (VA). The latter join to form the basilar artery (BA). Intracranially, these arteries anastomose around the circle of Willis and separate into the major intracranial vessels perfusing each hemisphere: the anterior cerebral artery (ACA), middle cerebral artery (MCA), and posterior cerebral artery (PCA). Pial vessel networks originate from the major intracranial vessels and are contiguous between cerebral vascular territories. However, their blood supply originates from their respective cranial arteries, and they predominantly supply a perfusion territory corresponding to that artery and its branches (illustrated by color-coding)

atherosclerosis, involve larger arterial vessels. There are also cerebrovascular pathologies such as hypertension that affect the entire circulation system. Nevertheless, regardless of the pathology and which region of the cerebrovascular system is affected, the consequences are seen in the distribution of the CBF at rest and in response to various stressors. Therefore, it is the measure of such parameters that enable the working back to the original perturbarion, as will be shown below.

# 6  Cerebrovascular Stress  Test

In various pathological conditions, there is adequate blood flow at rest to alay any symptoms. However, a reduced ability to compensate for localized reductions in blood flow due to an increase in upstream stenosis or a transient reduction in blood pressure may result in a transient loss of neurological function or, if prolonged, tissue infarction.

In the absence of symptoms can we test whether there are areas in the brain that have precarious blood flow and inadequate reserve flow when an increased resistance or blockage in upstream vessels develops or brain perfusion pressure falls?

The detection and characterization of collateral blood vessels and the assessment of the adequacy of their contribution to the maintenance of regional CBF regulation required the development of new techniques. Measures of cerebral vascular responsiveness, and the adequacy of both the collateral and main cerebrovascular circulations to provide CBF to the affected tissues, were needed [12, 34]. To this end, the vasculature needs to be stressed by the administration of a vasoactive stimulus and a measure of the response. A reduced regional CBF response to a vasodilatory stimulus, has been linked to increased stroke risk, and indeed considered to be the most important parameter for the diagnosis and management of patients with cerebrovascular disease [35].

## 7  Cerebrovascular Reactivity as the Brain Stress Test

Cerebrovascular reactivity (CVR) is a ratio of the change in CBF to the change in a vasoactive stimulus. It is therefore a measure of the regulatory ability of the cerebral vasculature and is used to assess cerebrovascular health with respect to its ability to compensate for changes in anatomy (e.g., stenosis) and other factors impacting the CBF supply (e.g., vasculitis). In recent years, carbon dioxide ($CO_2$) has been the most widely employed vasodilatory agent because of the ability to deploy it as a controlled and repeatable stimulus [36]. As the arterial partial pressure of $CO_2$ ($PaCO_2$) in the blood increases, the smooth muscles of the cerebral arterioles relax, decreasing the resistance to flow and increasing local CBF. Conversely, as the $PaCO_2$ decreases, the smooth muscle of the cerebral arterioles contract, increasing the resistance to flow and decreasing CBF. CVR is a surrogate test of the cerebral vasculature's responsiveness to the normal physiological demands such as (a) to meet increased metabolic demand via neurovascular coupling, (b) compensate for reductions in arterial blood pressure and decreases in oxygen delivery due to anemia or hypoxia. In short, CVR is a multimodle brain vasculature stress test.

In practice, the change in CBF with changes in $PaCO_2$ can be by using blood oxygen level dependent (BOLD) signal as a surrogate. BOLD changes in response to changes in $PaCO_2$ can be color coded and mapped voxel-by-voxel onto anatomical scans to generate CVR maps as shown in Fig. 3. These CVR maps show regions with a robust increase in BOLD signal in response to an increase in $PaCO_2$ (color-coded yellow to red) as well as unhealthy regions

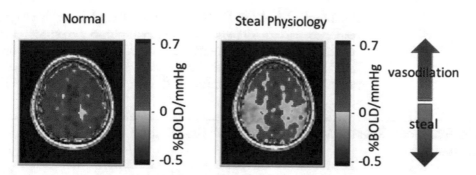

**Fig. 3** Axial slice of the CVR maps for (**a**) healthy subject and (**b**) individual with right middle cerebral artery occlusion with both hemispheres showing decrease in BOLD signal in response to increases in $PaCO_2$ representative of cerebrovascular steal. The CVR color scale is measured in %BOLD change per change in mmHg of $PaCO_2$ indicating the direction and magnitude of the change

where BOLD signal decreases with increases in $PaCO_2$ (color-coded light to dark blue). The latter regions are evidence for a phenomenon called 'cerebrovascular steal', so called because their blood flow has been 'stolen' from, and then redistributed to regions with a robust vasodilatory response.

## 8 The CBF Response to $PaCO_2$

Cerebrovascular resistance responds to $PaCO_2$ [17, 37], independently of CBF physiological regulatory mechanisms, by directly altering intravascular pH and cerebral vessel diameter. $PaCO_2$ changes the extracellular hydrogen ion concentration ($[H^+]$) environment of cerebrovascular smooth muscle [38–40], which opens potassium channels [41–43] decreasing smooth muscle membrane potential and hence tone and vessel diameter.

Therefore, in order for the autoregulation and neurovascular coupling, to function, the smooth muscle extracellular $[H^+]$ environment must be well maintained. However, the regulation of the interstitial $[H^+]$ environment depends on a number of factors [44], including the $PaCO_2$ and the local ionic environment [45]. Changes in $PaCO_2$ override the local $[H^+]$ regulation to produce $[H^+]$ changes that cause vasodilation and vasoconstriction. Similarly, hypoxia may act to degrade the local regulation of $[H^+]$ leading to an increase in $[H^+]$, vasodilation, and ultimately an increase in CBF. Under normal circumstances, $PaCO_2$ is under close regulation by the respiratory chemoreflex control system [46].

During CVR testing, changes in $PaCO_2$ are used as global vasoactive stimuli. In healthy subjects, the balanced changes in regional vascular resistances result in a symmetrical, stereotypical sigmoidal pattern of progressive increase in blood flow throughout the brain [17, 47–49]. It should be noted that this sigmoidal response pattern depends on arterial blood pressure remaining constant [50]. A rise in perfusion pressure constitutes a confounding factor to the effects of the vasoactive stimulus. Therefore blood pressure needs to be monitored during CVR testing [17, 51].

## 9    The Response Spectrum: from Normal CBF Distribution to the "Steal" Phenomenon

The balanced changes in CBF distribution seen in healthy subjects are not necessarily observed in patients with localized cerebral vascular disease. Where there are single or multifocal regions of vascular stenosis, due to, for example, atherosclerosis or vasculitis, the normally sigmoidal shaped changes in CBF with a progressive hypercapnic stimulus become unbalanced by asymetric changes in the local perfusion pressures and resistances.

Hypercapnia results in reductions in vascular resistance in all vessels. Reductions in perfusion pressure with hypercapnia occurs because cerebral inflow is limited to a level below CBF capacity in the parenchyma as a result of the high resistance of the extracranial cerebral arterial supply system, [52]. Consequently, when there is asymetric reductions in vascular resistance resulting from increases in $PaCO_2$ the vessels with the largest reductions in resistance draw the most blood, reducing the perfusion pressure, and consequently the flow to those beds that cannot match the reduction in resistance, a phenomenon labeled the cerebrovascular 'steal' [47, 53–55]. The redistribution aspect of the cerebral circulation resulting from asymetrical resistance of the extracranial cerebral arterial supply system can be modeled as a two-bed competitive balance of blood flow between co-perfused vascular regions (Fig. 4). Increases in flow in a region with a large vasodilatory reserve occur at the expense of one with less vasodilatory reserve.

This simple model of the blood flow distribution between vascular beds perfused in parallel shows that the CBF in each branch depends on both of their reductions in flow resistance and the shared perfusion pressure change [47]. In healthy people, the changes in flow resistance in the presence of a generalized vasodilatory stimulus (such as an increase in $PaCO_2$) are finely balanced between vascular regions throughout the brain, exhibiting a sigmoidal pattern of flow response [56]. However, in the presence of pathophysiology, the sigmoidal relationship as a function of the $PaCO_2$ may be distorted from that in health in terms of average slope, amplitude, or shifted to the right on the abcissa [57]. As a result, when viewed over a range of $PaCO_2$ stimulus, the flow response patterns may deviate from a sigmoidal response pattern and take on any of four patterns. These may be described as

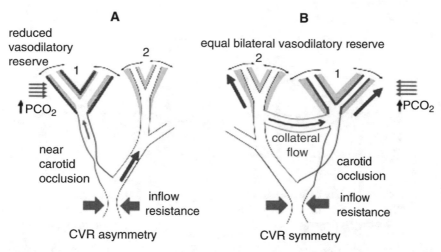

**Fig. 4** Vasodilatory reserve and collateral blood flow. Theoretical models of a brain vascular territory with a stenosed vessel branch (1) and a healthy branch (2) in parallel supplied by a major supply artery with high resistance. Their vascular reserves are shown shaded. (**a**) In order to supply the metabolic requirements of the tissue served by the stenosed vessel, the diameter of the vessels distal to the stenosis increase from their original diameter and encroach on their vasodilatory reserve. A vasodilatory stimulus such as an increase in $PaCO_2$ will stimulate all vessels to dilate. The healthy branch has unused vasodilatory reserve, reducing its resistance, incrdeasing its flow, resulting in a reduction of the perfusion pressure to both branches. Those vessels that have already encroached on their reserve, have limited capacity to further reduce their resistance, so the flow to these branches falls with the reduction in perfusion pressure. (**b**) With a collateral connection enabling flow past its stenosis, the stenosed branch need not encroach on its vasodilatory reserve. A vasodilatory stimulus will stimulate all vessels to dilate, and in this case, both branches have adequate vasodilatory reserve and share in the increased low; CVR is maintained in both branches

increasing, decreasing, biphasic concave up, and biphasic concave down [58], with only the increasing pattern (the normal response) having a sigmoidal shape (Figs. 5 and 6).

In summary, the redistribution of blood flow in the presence of a general vasodilatory stimulus is a sensitive indicator of functionally hemodynamic significant lesions. It seems CVR would be more sensitive and specific than a measure of the degree of stenosis or the resting CBF in identifying the hemodynamic significance of a vascular lesion to asses the risk of stroke; the latter fail to take into account the recruitable collateral blood flow revealed by CVR. Reductions in vasodilatory reserve will be identifyable by the generation of paradoxical intracerebral steal. The various patterns of CBF response to a ramp $CO_2$ challenge serve to identify types of pathology such as reduced vasodilatory reserve, absent reserve, and reduction in vasodilatory sensitivity to $PCO_2$. Since local flow changes in response to a stimulus are dependent on the patterns of resistance changes throughout the whole vascular tree, a more comprehensive method of CVR analysis is required. We suggest that of all our analyses, that of resistance changes over a range of $PaCO_2$ can address local pathophysiology.

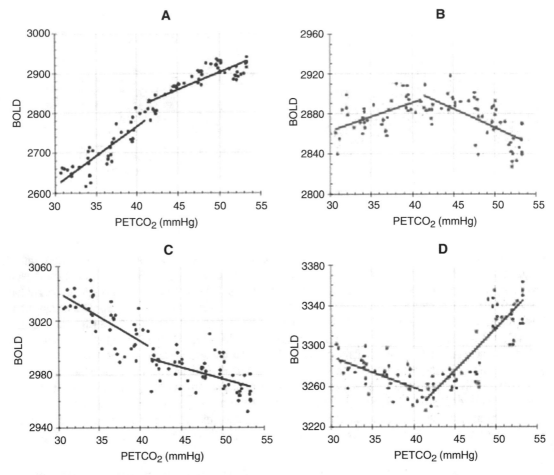

**Fig. 5** Examples of the patterns of flow response in individual voxels to a ramp increase in PetCO$_2$ from 30 to 55 mmHg. (**a**) A pattern best described as sigmoidal or approximated via a linear regression. (**b**) Biphasic response concave down. CBF initially increases with PaCO$_2$ then declines due to steal. (**c**) CBF declines as PaCO$_2$ increases over the whole PaCO$_2$ range (**d**) Biphasic concave up. Steal occurs during the initial rises in PaCO$_2$ from low levels then develops positive response with progressive hypercapnia. It is as if the sensitivity of dilation to CO$_2$ is shifted to the right

## 10  Measuring the Speed of Response

The CVR measures discussed so far provide an estimate of the amplitude and pattern of the cerebral vasculature's response to a vasodilatory stimulus. A measure of the speed of the response, whether rapid or slow, is a valuable dimension to describe vascular health. The speed of response arises from two factors: a blood transit time delay and a vascular resistance response time. By aligning the CO$_2$ and BOLD signals, the differences due to time delay between regions are minimized (assuming them to be less than the TR sampling period) leaving a measure of the speed of response. The BOLD response to the CO$_2$ stimulus can be analyzed to estimate the speed of response in the time or frequency domains.

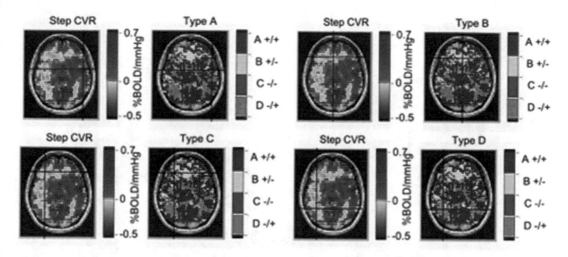

**Fig. 6** Examples of type maps as shown in Fig. 5 (previous) in an individual with bilateral moyamoya disease

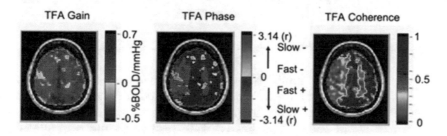

**Fig. 7** Transfer function analysis (TFA) maps from the same patient as found in Fig. 6 The TFA gain map provides another measure of CVR. The TFA phase map provides a measure of the speed of response. The TFA coherence map provides a measure of the consistency of the relation between the stimulus and response over time

In the time domain, the stimulus is convolved with a mono-exponential "dispersion" function and the time constant (tau) altered to provide the best fit with the BOLD response [59]. Tau therefore provides a measure of the speed of response. In the frequency domain, transfer function analysis (TFA) can be used to provide not only an estimate of the magnitude of the response, as CVR does, but also insight into the response dynamics [60]. The latter is provided by the phase relationship between the response and the stimulus as shown in Fig. 7 [61].

## 11   The Vasoactive Challenge

While various vasoactive stimuli may be used to challenge the cerebral vasculature, $CO_2$ remains closest to the ideal stimulus with the ability to present a standard challenge that is repeatable across and within subjects [36]. The best profile of the $CO_2$ stimulus may vary with the purpose of the challenge. Box-car or square

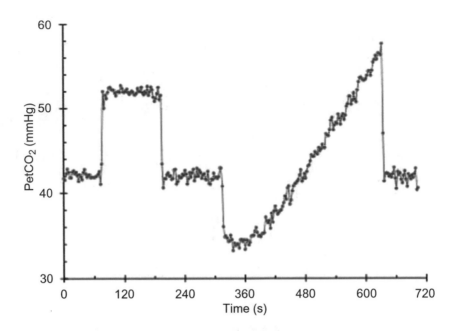

**Fig. 8** The $CO_2$ stimulus pattern. End-tidal $PCO_2$ vs. resamples to a TR = 2.4 s

wave stimuli [62] and sine wave stimuli [61] have been used to measure CVR. We have depended for most of our work on the pattern shown in Fig. 8, a combined step and ramp. The sequence of $PaCO_2$ changes is made relative to the resting $PaCO_2$ value, based on the idea that individuals control their CBF to provide the optimum regulation at the resting $PaCO_2$. The step change of $PaCO_2$ is set at 10 mmHg, a value designed to reveal abnormalities [47] without undue respiratory distress. Surveys have shown that tolerance is excellent for this step change [63]. Although the ramp stimulus eventually reaches higher values of $PaCO_2$, it does so slowly enough and lingers only momentarily at high $PCO_2$ so that tolerance is often better than the step change.

Both the step and ramp responses are used to determine CVR. The step response can also be analyzed to determine tau. The size of the step is 10 mmHg $PCO_2$ above resting [47] in order to provide a sufficient challenge. The ramp is preceded by a period of hyperventilation to lower $PaCO_2$ by about 10 mmHg below resting and continues to about 15 mmHg above resting so that the full range of vasoconstriction to vasodilation can be observed. The portion of the ramp from resting $PaCO_2$ to 10 mmHg above yields a CVR similar to that of the step. However, the step CVR is decreased in comparison to the ramp because it is affected by the speed of response (unless only steady-state measures are used). Speed of response is assumed to not affect the ramp CVR amplitude because the rise in $PaCO_2$ is slow enough to allow the slower responding voxels to respond fully.

## 12    Analyzing the BOLD Responses to PaCO$_2$

The step portion of the sequence is analyzed in two ways. First, a step CVR is calculated as the slope of $\Delta$BOLD vs. $\Delta$PETCO$_2$. Second, the BOLD response is analyzed to determine the speed of response as measured by the time constant (tau) of the assumed exponential changes. These measures generate step CVR and tau maps. In addition, the ramp portion between resting PaCO$_2$ and 'resting plus 10 mmHg' is fitted with linear regression to provide the slope of $\Delta$BOLD vs. $\Delta$PETCO$_2$. This 'ramp-resting + 10' CVR is therefore over the same PaCO$_2$ range as the step CVR and as mentioned previously is often greater than the step CVR because the speed of response factor is absent. The resulting step, 'ramp-resting + 10', and tau maps for the initial analysis are shown in Fig. 9.

## 13    Determining Abnormality

Because the CVR differs between regions, it is not immediately apparent whether the CVR in a particular location falls within the normal range. To overcome this ambiguity, a population of healthy individuals is tested. With all maps coregistered to the same standard space, the mean and standard deviation (SD) of each voxel can be calculated to provide an atlas of the normal range of CVR for all voxels. These normal values and their SD can then be used to classify each voxel in terms of SD, called $z$-scores [64]. Each voxel $z$-score can be color-coded and displayed as $z$-maps showing the deviation from healthy in SD units. An example of $z$-map is shown in Fig. 10.

A similar approach can be used to display abnormalities in tau. Resorting to an atlas can indicate whether a test-test difference is out of the normal range. Healthy subjects can be scanned on two occasions separated by weeks or months, and the CVR for each

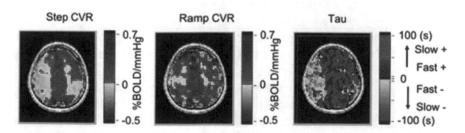

**Fig. 9** Example of CVR maps and corresponding speed of response (tau) map. Patient: a 56-year-old male with left-sided weakness and right middle cerebral artery occlusion. From left to right, step CVR, 'ramp-resting + 10' CVR, and tau. Note, in A, the large area of steal on ipsilateral side as well as a smaller area on the left parietal. With the ramp, there is considerably less steal seen, as the slower rise in PCO$_2$ provided time for the blood vessels to dilate. The tau map shows the area with the slowed response

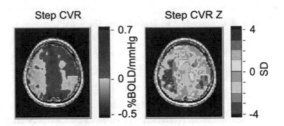

**Fig. 10** An example of CVR and corresponding $z$-map of the same patient found in Fig. 9. The $z$-map displays the extent that regions differ from the normal average atlas. Another perspective is that the z scores indicate the probability the voxel is within the normal range

voxel can be subtracted. This gives normative values for differences in CVR measures over time in all voxels. If then patients are scanned over time and the CVR for each voxel is subtracted, the differences can be compared voxelwise to those in a CVR difference atlas to identify which voxels fall within the normal range of those differences [65]. This tool can be used when following the natural evolution of a patient's disease or observing a response to a therapeutic intervention.

## 14    Cerebrovascular Resistance

The 'two vascular bed' model shown in Fig. 4 can be restated in terms of the flow resistance of each branch, analogous to an parallel electrical circuit model [57]. This model can be used to examine the ability of an individual region (e.g., voxel) to compete with a region with a robust vasodilatory reserve. By choosing a standard reference resistance response for the robust competition, the resistance responses of any voxels examined are comparable to each other (Fig. 11a). Taking the examined voxel BOLD change in response to $CO_2$ as a flow response and assuming values for the perfusion pressure and arterial resistance, the examined voxel resistance response can be calculated from the model. Considering that the resistance change over a progressive change in $PaCO_2$ is sigmoidal, it is not surprising to learn that the four patterns of flow response shown in Fig. 5 become sigmoidal patterns of resistance change when the model is solved.

Maps of these resistance response parameters (Fig. 6) provide detailed insight into the physiology and pathophysiology of vascular regulation in different brain regions that are not available from fitting the BOLD (i.e., flow) responses to $PaCO_2$. Indeed, the resistance response sigmoids may better describe the innate flow regulatory abilities of a region, un-confounded by the commmplex network of flow and perfusion changes resulting from the use of a global $CO_2$ stimulus. Figure 12 illustrates the resistance

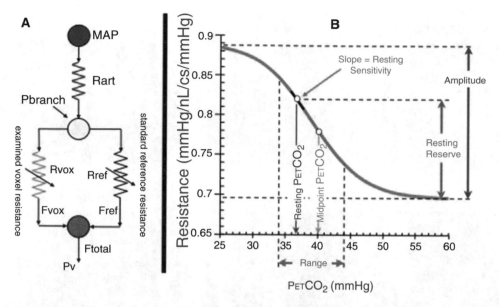

**Fig. 11** (**a**) The two-branch vascular resistance model used to convert voxel BOLD vs. $P_{ET}CO_2$ to voxel resistance vs. $P_{ET}CO_2$. (**b**) The sigmoidal change of vascular resistance as a function of $P_{ET}CO_2$ showing the various sigmoid parameters. *MAP* mean arterial pressure, *Rart* major cerebral arteries resistance, *Pbranch* voxel perfusion pressure, *Rvox* examined voxel resistance, *Rref* reference voxel resistance, *Fvox* examined voxel blood flow, *Fref* reference voxel blood flow, *Ftotal* total blood flow through both branches, *Pv* venous pressure

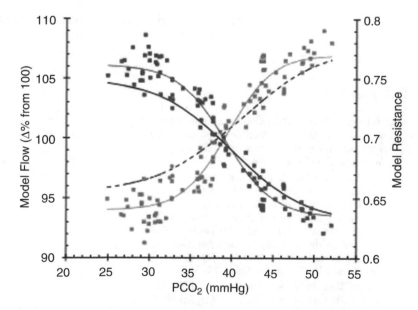

**Fig. 12** Deriving the resistance response to changes in $PaCO_2$ from the BOLD response to $PaCO_2$ in an example of voxel. The voxel BOLD signal (pink squares) is converted to % of the mean plus 100 to yield a model flow signal. The model reference resistance and the model flow points are then used to calculate the model resistance for each point (blue squares). A sigmoid is fitted to these voxel resistance points to characterize the voxel resistance response to $PaCO_2$ (solid blue line)

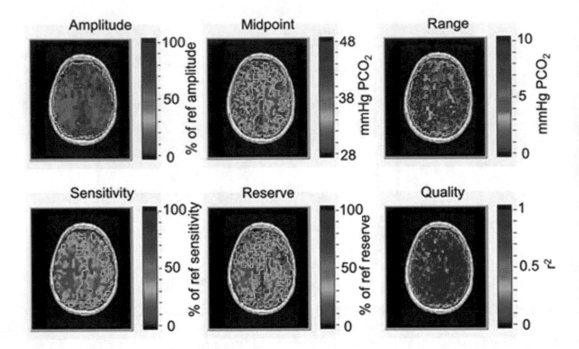

**Fig. 13** Resistance maps. Maps of sigmoid parameters as defined in Fig. 11b of the same patient found in Fig. 9. Amplitude, sensitivity, and reserve are scaled as % reference sigmoid. Midpoint is scaled as the $PCO_2$ (mmHg) and range as the $PCO_2$ span (mmHg). Quality is scaled as the $r^2$ of the resistance sigmoid fit

changes in a voxel in response to a ramp $CO_2$ challenge as derived from the BOLD/flow response, and the corresponding resistance parameter maps of these responses are illustrated in Fig. 13.

## References

1. Helenius J et al (2003) Cerebral hemodynamics in a healthy population measured by dynamic susceptibility contrast MR imaging. Acta Radiol 44(5):538–546
2. Ito H et al (2004) Database of normal human cerebral blood flow, cerebral blood volume, cerebral oxygen extraction fraction and cerebral metabolic rate of oxygen measured by positron emission tomography with 15O-labelled carbon dioxide or water, carbon monoxide and oxygen: a multicentre study in Japan. Eur J Nucl Med Mol Imaging 31(5):635–643
3. Willie CK et al (2014) Integrative regulation of human brain blood flow. J Physiol 592 (5):841–859
4. Nishimura N et al (2007) Penetrating arterioles are a bottleneck in the perfusion of neocortex. Proc Natl Acad Sci U S A 104 (1):365–370
5. Hall CN et al (2014) Capillary pericytes regulate cerebral blood flow in health and disease. Nature 508(7494):55–60
6. Attwell D et al (2016) What is a pericyte? J Cereb Blood Flow Metab 36(2):451–455
7. Tan CO, Taylor JA (2014) Integrative physiological and computational approaches to understand autonomic control of cerebral autoregulation. Exp Physiol 99(1):3–15
8. Tzeng Y-C et al (2014) Fundamental relationships between blood pressure and cerebral blood flow in humans. J Appl Physiol (1985) 117(9):1037
9. Attwell D et al (2011) Glial and neuronal control of brain blood flow. Nature 468 (7321):232–243
10. Phillips AA et al (2016) Neurovascular coupling in humans: physiology, methodological advances and clinical implications. J Cereb Blood Flow Metab 36(4):647–664

11. Zarrinkoob L et al (2015) Blood flow distribution in cerebral arteries. J Cereb Blood Flow Metab 35(4):648–654

12. Liebeskind DS (2003) Collateral circulation. Stroke 34(9):2279–2284

13. Moody DM, Bell MA, Challa VR (1990) Features of the cerebral vascular pattern that predict vulnerability to perfusion or oxygenation deficiency: an anatomic study. AJNR Am J Neuroradiol 11(3):431–439

14. Willie CK et al (2012) Regional brain blood flow in man during acute changes in arterial blood gases. J Physiol 590(14):3261–3275

15. Mardimae A et al (2012) The interaction of carbon dioxide and hypoxia in the control of cerebral blood flow. Pflugers Arch 464 (4):345–351

16. Cohen PJ et al (1967) Effects of hypoxia and normocarbia on cerebral blood flow and metabolism in conscious man. J Appl Physiol 23(2):183–189

17. Battisti-Charbonney A, Fisher J, Duffin J (2011) The cerebrovascular response to carbon dioxide in humans. J Physiol 589 (Pt 12):3039–3048

18. Borzage MT et al (2016) Predictors of cerebral blood flow in patients with and without anemia. J Appl Physiol 120(8):976–981

19. Duffin J, Hare GMT, Fisher JA (2020) A mathematical model of cerebral blood flow control in anaemia and hypoxia. J Physiol 598 (4):717–730

20. Brown MM, Wade JP, Marshall J (1985) Fundamental importance of arterial oxygen content in the regulation of cerebral blood flow in man. Brain 108(Pt 1):81–93

21. Bush AM et al (2016) Determinants of resting cerebral blood flow in sickle cell disease. Am J Hematol 91(9):912–917

22. Wolff CB (2000) Cerebral blood flow and oxygen delivery at high altitude. High Alt Med Biol 1(1):33–38

23. Powell FL, Fu Z (2008) HIF-1 and ventilatory acclimatization to chronic hypoxia. Respir Physiol Neurobiol 164(1–2):282–287

24. Poellinger L, Johnson RS (2004) HIF-1 and hypoxic response: the plot thickens. Curr Opin Genet Dev 14(1):81–85

25. Hulbert ML et al (2017) Normalization of cerebral hemodynamics after hematopoietic stem cell transplant in children with sickle cell anemia. Blood 130(Suppl 1):2245–2245

26. Victor M, Ropper AH, Adams R (2000) Adams and Victor's principles of neurology, 7th edn. McGraw-Hill Professional, New York, NY

27. Fierstra J et al (2010) Steal physiology is spatially associated with cortical thinning. J Neurol Neurosurg Psychiatry 81(3):290–293

28. Sam K et al (2016) Impaired dynamic cerebrovascular response to hypercapnia predicts development of white matter hyperintensities. NeuroImage: Clin 11:796–801

29. Sam K et al (2016) Development of white matter hyperintensity is preceded by reduced cerebrovascular reactivity. Ann Neurol 80:277

30. Balucani C et al (2012) Cerebral hemodynamics and cognitive performance in bilateral asymptomatic carotid stenosis. Neurology 79 (17):1788–1795

31. Ben Hassen W et al (2019) Inter- and intraobserver reliability for angiographic leptomeningeal collateral flow assessment by the American Society of Interventional and Therapeutic Neuroradiology/Society of Interventional Radiology (ASITN/SIR) scale. J Neurointerv Surg 11(4):338–341

32. Lima FO et al (2010) The pattern of leptomeningeal collaterals on CT angiography is a strong predictor of long-term functional outcome in stroke patients with large vessel intracranial occlusion. Stroke 41(10):2316–2322

33. Cipolla MJ (2010) The cerebral circulation. Morgan & Claypool, San Rafael, CA

34. Powers WJ (1991) Cerebral hemodynamics in ischemic cerebrovascular disease. Ann Neurol 29(3):231–240

35. Reinhard M et al (2014) Cerebrovascular reactivity predicts stroke in high-grade carotid artery disease. Neurology 83(16):1424–1431

36. Fierstra J et al (2013) Measuring cerebrovascular reactivity: what stimulus to use? J Physiol 591(Pt 23):5809–5821

37. Reivich M (1964) Arterial Pco2 and cerebral hemodynamics. Am J Phys 206:25–35

38. Lassen NA (1968) Brain extracellular pH: the main factor controlling cerebral blood flow. Scand J Clin Lab Invest 22(4):247–251

39. Murkin JM (2007) Cerebral autoregulation: the role of CO2 in metabolic homeostasis. Semin Cardiothorac Vasc Anesth 11 (4):269–273

40. Kontos HA et al (1977) Local mechanism of CO2 action of cat pial arterioles. Stroke 8 (2):226–229

41. Wei EP, Kontos HA (1999) Blockade of ATP-sensitive potassium channels in cerebral arterioles inhibits vasoconstriction from hypocapnic alkalosis in cats. Stroke 30(4):851–853. discussion 854

42. Longden TA, Nelson MT (2015) Vascular inward rectifier K+ channels as external K+

sensors in the control of cerebral blood flow. Microcirculation 22(3):183–196

43. Rasmussen JK, Boedtkjer E (2018) Carbonic anhydrase inhibitors modify intracellular pH transients and contractions of rat middle cerebral arteries during $CO_2/HCO_3(-)$ fluctuations. J Cereb Blood Flow Metab 38 (3):492–505

44. Chesler M (2003) Regulation and modulation of pH in the brain. Physiol Rev 83 (4):1183–1221

45. Hughes R, Brain MJ (2013) A simplified bedside approach to acid–base: fluid physiology utilizing classical and physicochemical approaches. Anaesth Inten Care Med 14 (10):445–452

46. Duffin J (2005) Role of acid-base balance in the chemoreflex control of breathing. J Appl Physiol 99(6):2255–2265

47. Sobczyk O et al (2014) A conceptual model for CO2-induced redistribution of cerebral blood flow with experimental confirmation using BOLD MRI. NeuroImage 92:56–68

48. Bhogal A et al (2014) Investigating the non-linearity of the BOLD cerebrovascular reactivity response to targeted hypo/hypercapnia at 7T. NeuroImage 98:296

49. Bhogal AA et al (2016) The BOLD cerebrovascular reactivity response to progressive hypercapnia in young and elderly. NeuroImage 139:94–102

50. Regan RE, Fisher JA, Duffin J (2014) Factors affecting the determination of cerebrovascular reactivity. Brain Behav 4(5):775–788

51. Fan J-L et al (2015) AltitudeOmics: resetting of cerebrovascular CO2 reactivity following acclimatization to high altitude. Front Physiol 6:394

52. Faraci F, Heistad D (1990) Regulation of large cerebral arteries and cerebral microvascular pressure. Circ Res 66(1):8–17

53. Brawley BW (1968) The pathophysiology of intracerebral steal following carbon dioxide inhalation, an experimental study. Scand J Clin Lab Investig Suppl XIII:B:102

54. Symon L (1968) Experimental evidence for "intracerebral steal" following CO2 inhalation. Scand J Clin Lab Investig Suppl XIII:A:102

55. Conklin J et al (2010) Impaired cerebrovascular reactivity with steal phenomenon is associated with increased diffusion in white matter of patients with Moyamoya disease. Stroke 41 (8):1610–1616

56. Bhogal AA et al (2015) Examining the regional and cerebral depth-dependent BOLD cerebrovascular reactivity response at 7T. NeuroImage 114:239–248

57. Duffin J et al (2018) Cerebrovascular resistance: the basis of cerebrovascular reactivity. Front Neurosci 12:409

58. Fisher JA et al (2017) Assessing cerebrovascular reactivity by the pattern of response to progressive hypercapnia. Hum Brain Mapp 38:3415

59. Poublanc J et al (2015) Measuring cerebrovascular reactivity: the dynamic response to a step hypercapnic stimulus. J Cereb Blood Flow Metab 35:1–11

60. Duffin J et al (2015) The dynamics of cerebrovascular reactivity shown with transfer function analysis. NeuroImage 114:207–216

61. Blockley NP et al (2011) An improved method for acquiring cerebrovascular reactivity maps. Magn Reson Med 65(5):1278–1286

62. Mandell DM et al (2008) Mapping cerebrovascular reactivity using blood oxygen level-dependent MRI in patients with arterial steno-occlusive disease: comparison with arterial spin labeling MRI. Stroke 39 (7):2021–2028

63. Spano VR et al (2013) CO2 blood oxygen level-dependent MR mapping of cerebrovascular reserve in a clinical population: safety, tolerability, and technical feasibility. Radiology 266 (2):592–598

64. Sobczyk O et al (2015) Assessing cerebrovascular reactivity abnormality by comparison to a reference atlas. J Cereb Blood Flow Metab 35 (2):213–220

65. Sobczyk O et al (2016) Identifying significant changes in cerebrovascular reactivity to carbon dioxide. AJNR Am J Neuroradiol 37 (5):818–824

# Chapter 2

# Experimental Protocols in CVR Mapping

## Marat Slessarev

## Abstract

Cerebrovascular reactivity (CVR) mapping requires a method that measures cerebral blood flow (CBF) and a vasoactive stimulus that induces a change in CBF. In this chapter, we focus on the vasoactive stimuli used in CVR mapping. Specifically, we review why optimizing vasoactive stimulus is necessary for CVR mapping, define characteristics of an optimal stimulus, and use these characteristics to critically appraise available vasoactive stimuli.

**Keywords** Cerebrovascular reactivity, $CO_2$, Rebreathing, Cerebral blood flow, Breath-holding

## 1 Introduction

Cerebrovascular reactivity (CVR) mapping combines methods that measure changes in cerebral blood flow (CBF) and induce a vasoactive stimulus that causes a change in CBF necessary for computation and mapping of CVR. Methods for measuring CBF vary in terms of their portability and spatial and temporal resolution and are reviewed in detail elsewhere [1]. In this chapter, we will review the methods used to induce vasoactive stimuli for the purpose of CVR measurement and mapping. First, we will discuss why optimizing vasoactive stimulus is important. Second, we will identify the characteristics of an optimal stimulus. Finally, we will review the different types of vasoactive stimuli that are available for measurement and mapping of CVR and critically appraise them using characteristics defined in the second section.

## 2 Why Is It Important to Optimize Vasoactive Stimulus in CVR Mapping?

The goal of CVR mapping is to assess the functional status of cerebrovascular circulation by probing its ability to control the distribution of CBF when exposed to a vasoactive stimulus. Given that the *direction*, *magnitude*, *temporal profile*, and *reproducibility*

Jean Chen and Jorn Fierstra (eds.), *Cerebrovascular Reactivity: Methodological Advances and Clinical Applications*, Neuromethods, vol. 175, https://doi.org/10.1007/978-1-0716-1763-2_2, © Springer Science+Business Media, LLC, part of Springer Nature 2022

of the vasoactive stimulus can affect CBF response and computation of CVR, it is critical to ensure that these variables can be precisely measured and optimally controlled to ensure accuracy in CVR mapping.

### 2.1 Direction of Vasoactive Stimulus from Baseline

Given that CBF response to a vasoactive stimuli may not be linear [2], the direction of the stimulus from baseline can affect the magnitude of the CBF response and, hence, CVR mapping. For example, changes in arterial partial pressure of $CO_2$ ($PaCO_2$) are often used as a vasoactive stimulus to measure CVR. Given that obtaining $PaCO_2$ requires arterial puncture or insertion of an invasive arterial line, most CVR mapping studies use end-tidal partial pressure of $CO_2$ ($PetCO_2$) as a surrogate to approximate $PaCO_2$. However, the relationship between CBF and $PetCO_2$ is nonlinear, with greater sensitivity of CBF to increase (hypercapnia) rather than decrease (hypocapnia) in $PetCO_2$ [3]. Furthermore, the CBF-$PetCO_2$ relationship displays hysteresis, with greater sensitivity when $PetCO_2$ is changed from hypo- to hypercapnia [3]. As a result, hypocapnic stimulus may result in blunted CVR response compared to hypercapnic stimulus [4, 5]. In addition to considering the direction of vasoactive stimulus from baseline, recent studies suggest that sigmoid rather than linear models of CBF-$PetCO_2$ relationship may result in more accurate CVR mapping [2, 5].

### 2.2 Magnitude of Vasoactive Stimulus

Improving signal to noise ratio (SNR) is the cornerstone of CVR mapping. When blood oxygen level-dependent (BOLD) MRI is used, $PetCO_2$ change of at least 2 mmHg has been suggested to enable adequate SNR for CVR assessment [6]. In this regard, some methods like acetazolamide administration may result in maximal vasodilatory stimulus, while other methods, such as breath-holding, may compromise SNR due to too small of a change in $PetCO_2$. Using both stimuli in the same participant may therefore result in drastically different CVR values. Furthermore, since CBF-$PetCO_2$ relationship has limits at extremes of hypo- and hypercapnia [7, 8] and given that these limits differ between individuals [7], using maximal vasodilatory stimulus like acetazolamide may underestimate CVR unless the model incorporates the sigmoid shape of CBF-$PetCO_2$ relationship into CVR calculation [2].

### 2.3 Temporal Profile of Vasoactive Stimulus

The temporal profiles of available vasoactive stimuli include ramp [9, 10], square-wave or block [11, 12], and complex (e.g., sinusoidal) [13] patterns. Methods such as breath-holding, administration of acetazolamide, rebreathing, and use of vasoactive medications to change blood pressure all result in ramp stimuli. The advantages of these methods are that they can be relatively easy to administer (e.g., breath-holding, administration of acetazolamide) and can be used to assess the shape [5] and limits [7] of CBF-$PetCO_2$ relationship (e.g., using rebreathing). However, the rate of change in the vasoactive stimulus varies between participants depending on factors such as rate of metabolic $CO_2$ production, lung volumes, and participant tolerance of hypercapnia. Furthermore, ramp methods are not suitable for CVR mapping that utilize CBF measurement methods that drift over time, such as BOLD MRI, or those that require steady-state conditions, such as arterial spin labeling (ASL) MRI [14]. Finally, ramp methods are not suitable for estimating temporal characteristics of CBF response to vasoactive stimuli, as slow rate of change in vasoactive stimulus is the rate-limiting factor [11]. As a result, square-wave or block patterns are preferred in CVR mapping that utilize BOLD and ASL MRI [15], as they mitigate BOLD signal drift and enable assessment of CVR dynamics. Complex sinusoidal patterns are a new frontier in CVR mapping that enable assessment of both the magnitude and dynamics of CVR [13, 15]. However, both square-wave and complex patterns require advanced methods to induce vasoactive stimuli, which increases complexity and costs.

**2.4  Reproducibility of Vasoactive Stimulus**

Interpretation of CVR maps includes comparison of study results with a reference standard (e.g., CVR atlas) [16], pre-intervention maps [17–19], or longitudinal follow-up over time [20–22]. In all three scenarios, it is critical that the vasoactive stimulus is reproducible both within and between participants to ensure that observed changes in CVR are not simply due to variance in stimulus direction, magnitude, or temporal pattern [23]. In this regard, methods such as prospective targeting or dynamic end-tidal forcing result in more reproducible stimuli than methods such as breath-holding or $CO_2$ inhalation.

**2.5  What Are the Characteristics of an Optimal Method for Inducing Vasoactive Stimulus in CVR Mapping?**

The optimal method for CVR mapping should enable precise control of the direction, magnitude, and temporal pattern of the vasoactive stimulus as well as ensure its reproducibility within and between participants. This would ensure that the CVR measurements actually represent changes in the states of health or disease, rather than variations due to differences in the applied vasoactive stimulus. Given that CVR mapping is often applied to clinical populations, including those with the vulnerable brain (e.g., concussion, cerebrovascular disease) [24, 25], the optimal method should be safe and noninvasive, have minimal side effects, and allow rapid reversal of the vasoactive stimulus in case of patient intolerance of adverse event.

## 3  Critical Appraisal of Available Vasoactive Stimuli

Vasoactive stimuli used in CVR mapping can be broadly divided into (a) manipulation of blood pressure or cardiac output and (b) manipulation of $CO_2$ (Table 1). While physiological studies attempt to separate the effects of blood pressure, cardiac output,

**Table 1**

**Comparison of the methods and respective vasoactive stimuli available for CVR mapping**

| Method | Stimulus | Magnitude | Temporal characteristics | Reproducibility | Safety and side effects |
|---|---|---|---|---|---|
| Methods that ↓ CBF | | | | | |
| Low body negative pressure | ↓ MAP ↓ CO | Wide range, titratable | Step, ramp, transient, and sustained | +++++ | Risk of syncope, ischemia |
| Cuff deflation | ↓ MAP | Medium range, not titratable | Step, transient | +++++ | Risk of ischemia |
| IV infusion of vasodilators | ↓ MAP | Wide range, titratable | Steady-state and ramp, transient, and sustained | +++++ | Invasive, risk of ischemia |
| Hyperventilation | ↓ $PetCO_2$ | Wide range, hard to titrate | Step and ramp; transient, hard to sustain | +++ | Dizziness, risk of ischemia |
| Methods that ↑ CBF | | | | | |
| IV infusion of vasopressors | ↑ MAP | Wide range, titratable | Steady-state and ramp, transient, and sustained | +++++ | Invasive, ischemia with extravasation |
| Breath-holding | ↑ $PetCO_2$ | Small range, not titratable | Variable; transient | + | Dizziness, dyspnea |
| Acetazolamide administration | ↑ $PbtCO_2$ | Wide range, not titratable | Ramp; sustained | +++ | Dizziness, headache, nausea, blurred vision, hard to reverse |
| $CO_2$ inhalation via face mask | ↑ $PetCO_2$ | Wide range, hard to titrate | Ramp; transient and sustained | +++ | Dizziness, headache, flushing, tachypnea, dyspnea |
| Methods that both ↑ and ↓ CBF | | | | | |
| Rebreathing | ↑ and ↓[a] $PetCO_2$ | Wide range, titratable | Ramp, sustained | ++++ | Dizziness, headache, flushing, tachypnea, dyspnea |
| Dynamic end-tidal forcing | ↑ and ↓[a] $PetCO_2$ | Wide range, titratable | Step, ramp, transient, and sustained | +++++ | Similar to hypocapnia and $CO_2$ inhalation |
| Prospective end-tidal targeting | ↑ and ↓[a] $PaCO_2$ | Wide range, titratable | Step, ramp, complex; transient and sustained | +++++ | Similar to hypocapnia and $CO_2$ inhalation |

*MAP* mean arterial pressure, *CO* cardiac output, *PetCO₂* end-tidal $PCO_2$, *PbtCO₂* brain tissue $PCO_2$, *PaCO₂* arterial $PCO_2$

[a]These methods incorporate hyperventilation to enable hypocapnic stimulus

and $CO_2$ on CBF, change in all of these variables likely interrogates the same cerebrovascular reserve, which is defined by the ability of cerebral vessels to regulate CBF by changing their diameter [23, 24, 26]. It therefore follows that changes in blood pressure, cardiac output, and $CO_2$ can be used as effective vasoactive stimuli to assess CVR. However, gradual reduction of perfusion pressure causes progressive vasodilation and dampening of CBF response to $CO_2$. In a classic experiment, Harper and Glass showed that reduction of mean arterial pressure (MAP) in dogs resulted in dampening and eventually complete abolition of CBF response to $CO_2$ [8]. The failure of cerebral vasculature to constrict in response to hypocapnia during hypotensive conditions suggests that maintaining perfusion pressure takes precedence over $CO_2$ reactivity [8]. Similar results were later confirmed in humans [27, 28]. Given that cerebrovascular disease affects perfusion pressure to specific areas of the brain, changes in $CO_2$ represent an ideal stimulus to unmask affected brain regions during CVR mapping [14]. In contrast, manipulation of blood pressure and cardiac output is more invasive and cumbersome and can lead to tissue ischemia and hypoxia by further reducing perfusion pressure. As a result, manipulation of $CO_2$ has generally emerged as a preferred method in CVR mapping studies. In this section, we will use criteria for optimal vasoactive stimulus described in the previous section to critically appraise available vasoactive stimuli that can be utilized in CVR mapping.

### 3.1  Low Body Negative Pressure

Lower body negative pressure causes a blood shift from upper to lower body compartments. By adjusting the level of low body negative pressure, one can produce precisely titratable reduction in blood pressure, cardiac output, and CBF from baseline [29], while adjusting the rate of change in low body negative pressure will influence the temporal pattern of the vasoactive stimuli. Prolonged periods of lower body negative pressure can lead to ischemia and presyncope, so despite its ability to precisely control the magnitude, temporal pattern, and reproducibility of the delivered vasoactive stimulus, this method is not suitable for clinical CVR mapping.

### 3.2  Thigh Cuff Deflation

Similar to lower body negative pressure, sudden release of an inflated thigh cuff induces a transient reduction in blood pressure (on average from 10% to 60% from baseline) that lasts approximately 10–30 s, followed by a return of blood pressure to baseline [30]. This method is relatively simple to use and has been shown to work in the MRI scanner. However, in contrast to low body negative pressure, which allows precise titration of the vasoactive stimulus, cuff deflation cannot be titrated, which limits its reproducibility between participants (i.e., the blood pressure response will vary between participants depending on their physiological state). Furthermore, cuff deflation only allows assessment of the response to transient reduction in blood pressure. Finally, while relatively safe in healthy participants, reduction in blood pressure in patients with vulnerable brain may results in ischemia and presyncope and is therefore not suitable for widespread clinical use.

### 3.3 Infusion of Vasoactive Medication

Infusion of vasopressors and vasodilators can be used to manipulate MAP. This approach is fairly invasive, as it requires insertion of an indwelling arterial catheter and continuous hemodynamic and cardiac monitoring to ensure participant safety. Titration of vasoactive agents enables precise and reproducible stimulus, although the temporal course is limited to ramp pattern. While ramp increase and reduction in MAP with vasoactive agents have been used to assess static autoregulation in healthy volunteers [31], the invasiveness of this method, its requirement for continuous hemodynamic monitoring, and potential risk of ischemia due to reduction of blood pressure make it less practical for routine clinical CVR mapping.

### 3.4 Hyperventilation

Hyperventilation, achieved by an increase in tidal volume and/or respiratory rate, leads to reduction in $PetCO_2$ and is a potent vasoconstrictive stimulus. With appropriate coaching and visual and auditory feedback, participants can learn to hyperventilate to a target $PetCO_2$, although variable tolerance of hypocapnia and respiratory fatigue may affect reproducibility in $PetCO_2$ stimulus between participants. For some participants, being placed in the MRI scanner induces anxiety and further hyperventilation, which again affects reproducibility of $PetCO_2$ stimulus. Hyperventilation allows both step and ramp reductions in $PetCO_2$, but ensuing vasoconstriction can induce ischemia in vulnerable patients and is generally avoided in favor of hypercapnic stimuli during CVR mapping [14].

### 3.5 Breath-Holding

Breath-holding increases $PetCO_2$ by temporarily reducing respiratory $CO_2$ elimination. The benefits of this approach include its noninvasiveness and simplicity. Given challenges with estimating $PetCO_2$ at the end of the breath-hold, early studies used breath-hold time as a surrogate indicator of the $PetCO_2$ stimulus [32–34]. However, while this approach is practically appealing, it does not represent an accurate estimation of changes in $PetCO_2$, which translates into inaccuracy with CVR mapping for the following reasons. First, the rate of rise in $PetCO_2$ during a breath-hold depends in part on the metabolic $CO_2$ production, which differs within and between participants. Similarly, the rate of rise in $PetCO_2$ also depends on the size of the respiratory reservoir, which in this case corresponds to the size of the lungs at the start of the breath-hold. At a given metabolic rate of $CO_2$ production, the $PetCO_2$ at the end of breath-hold will be higher in participants with smaller starting lung volumes. The lung volumes can vary between participants based on age, sex, height, obesity, and lung disease [35]. Within a given participant, lung volumes will differ depending on whether the breath-hold was initiated at the end of expiration or inspiration. Second, if all other factors remain constant, the final $PetCO_2$ at the end of the breath-hold will depend on the starting $PetCO_2$, which varies between participants. Finally, breath-hold time can vary widely both within and between participants [36]. To mitigate these shortcomings, more recent CVR studies attempted to standardize methodology by initiating breath-hold at the end of expiration and measuring $PetCO_2$ values prior to and at completion of the breath-hold [37]. However, the differences in breath-hold time, lung size, starting $PetCO_2$, and metabolic $CO_2$ production between participants cannot be standardized, which limits the precision, magnitude, temporal characteristics, and reproducibility of this vasodilatory stimulus. As a result, despite its simplicity and safety, breath-holding remains an unsuitable method for quantitative clinical CVR mapping.

**3.6 Acetazolamide Administration**

Intravenous administration of carbonic anhydrase inhibitor acetazolamide increases brain tissue partial pressure of $CO_2$ ($PbtCO_2$), which results in vasodilation and CBF increase. The standard dose used in most studies is 1 g, which leads to maximal vasodilatory response [28]. While adjusting the dose of acetazolamide may produce sub-maximal vasodilation, precise titration of vasodilatory stimulus is challenging given the wide variability between participants in drug pharmacokinetics and pharmacodynamics. Furthermore, the increase in $PbtCO_2$ induces ventilation and increases $CO_2$ elimination, which can lead to reduction in $PetCO_2$ and primary vasodilatory stimulus. As a result, the temporal pattern of vasodilatory stimulus following acetazolamide administration is unpredictable. Despite its simplicity, ease of administration, and relative safety, acetazolamide administration has fallen out of favor in CVR mapping protocols due to lack of control over the stimulus magnitude, temporal pattern, and reproducibility, as well as high prevalence of side effects such as dizziness, headache, nausea, and other neurological symptoms [38]. Furthermore, given the relatively long half-life of elimination for acetazolamide (4–8 h) [39], its adverse effects can last for a long time and cannot be reversed.

**3.7 $CO_2$ Inhalation via a Face Mask**

Adding $CO_2$ to the inspired gas can increase $PetCO_2$ and serve as a vasodilatory stimulus in CVR mapping. This method is logistically simple and safe and requires little cooperation from the participants. However, fixed inspired $CO_2$ concentration does not result in reproducible vasodilatory stimulus for the following reasons. First, inhalation of $CO_2$ activates respiratory chemoreflex, which increases minute ventilation and $CO_2$ elimination and works against increasing $PaCO_2$ [40]. Given that chemoreflex sensitivity to $CO_2$ varies between individuals [41], the final $PaCO_2$ will also differ between participants despite similar inspired $CO_2$ fraction and, in participants with high chemoreflex sensitivity, can be the same or lower than baseline $PaCO_2$ [42, 43]. Hyperventilation (e.g., due to anxiety or claustrophobia inside MRI scanner) can further counter the increase in $PaCO_2$ during $CO_2$ inhalation [43]. Second, an increase in ventilation during $CO_2$ inhalation increases end-tidal partial pressure of $O_2$ ($PetO_2$), which can contribute to noise if BOLD signal is used as a measure of CBF, contaminating as much as 9–17% of the total signal [44]. This detrimental effect of $PetO_2$ on the BOLD SNR is worse if $CO_2$-oxygen instead of $CO_2$-air gas mixture is used during $CO_2$ inhalation. Finally, while $PetCO_2$ is often used as an indicator of the vasoactive stimulus, CBF actually responds to changes in $PaCO_2$ [45]. In healthy participants, there is a small yet persistent gradient between $PetCO_2$ and $PaCO_2$ [46]. This gradient can vary within participants with exercise [47] and body position [48] and between participants with age [49] and disease [50]. In summary, while $CO_2$ inhalation can induce wide range of $PetCO_2$ with some control over temporal patterns, it results in highly variable $PetCO_2$ within and between participants and is therefore does not represent an optimal vasoactive stimulus for quantitative CVR mapping.

## 3.8 Rebreathing

Rebreathing of exhaled gas leads to a gradual increase in $PetCO_2$ and has been used in CVR studies. It is safe, noninvasive, and relatively easy to implement. Similar to breath-holding, the rate of rise in $PetCO_2$ depends on the metabolic rate of $CO_2$ production and the size of the rebreathing reservoir. However, with rebreathing, the reservoir consists of the lungs plus the rebreathing bag, such that initiation of rebreathing at end-expiration and smaller bag size leads to faster rate of rise in $PetCO_2$. Owing to the slow rate of metabolic $CO_2$ production, rebreathing only enables ramp temporal pattern. As a result, it is not suitable for studying CVR dynamics and is not ideal for use with CBF MRI methods that drift over time (e.g., BOLD) or require steady states (e.g., ASL) [14]. In contrast to $CO_2$ inhalation, rebreathing eliminates $PetCO_2$ to $PaCO_2$ gradient, so that the measured change in $PetCO_2$ represents an actual vasodilatory stimulus. Because rebreathing leads to metabolic $O_2$ consumption, servo-control systems are used to maintain constant $O_2$ levels in the reservoir. Duffin's modified rebreathing uses hyperventilation to achieve lower starting $PetCO_2$ prior to initiation of rebreathing and enables assessment of CVR over a wide range of hypo- and hypercapnia [51]. It has been used to establish the limits of CBF response to $CO_2$ and define the cerebrovascular reserve [7, 52]. Although starting and final $PetCO_2$ during rebreathing vary between participants based on their ability to tolerate hyperventilation and elevated $CO_2$ levels, the wide $PetCO_2$ range ensures that this method is fairly reproducible within and between participants.

**3.9 Dynamic
End-Tidal Forcing**

Dynamic end-tidal forcing is a servo-control method that was developed to study ventilatory and CBF responses to changes in $CO_2$ and $O_2$ [3, 53, 54]. It utilizes a computer-controlled gas blender that mixes source pressurized gases to achieve an inspiratory concentration of $CO_2$ and $O_2$ that is estimated to achieve desired target PetCO$_2$ and PetO$_2$ levels. Actual PetCO$_2$ and PetO$_2$ levels are measured from the participants, and a feedback algorithm then adjusts inspired $CO_2$ and $O_2$ to "force" the PetCO$_2$ and PetO$_2$ toward desired targets [55]. To achieve ramp and step transitions between PetCO$_2$ and PetO$_2$ target, the algorithm adjusts inspired $CO_2$ and $O_2$ concentrations and uses "overpressure" algorithm to achieve step patterns of change in PetCO$_2$. Dynamic end-tidal forcing can induce a wide range of precisely titratable PetCO$_2$ levels, enables various time profiles of changes in PetCO$_2$, and is reproducible within and between participants [56]. However, its utility for clinical CVR mapping remains limited for the following reasons. First, it requires high gas flows (50–70 L/min) via large-bore tubing in order to match participant's peak inspiratory flows. Generation of such high flows requires large source gas tanks and systems to humidify inspired gas. Second, like any feedback system, it is prone to instability, especially if analysis of expired gas is delayed (e.g., in the context of MRI CVR mapping that has to utilize long extension lines between MRI core and control room). Third, similar to inhalation of $CO_2$ mixtures, there is persistent gradient between PetCO$_2$ and PaCO$_2$, resulting in imprecision of the vasoactive stimulus. Fourth, the source gases contain anoxic mixtures (i.e., pure nitrogen and $CO_2$), so if there is failure of $O_2$ delivery, there is a serious risk of severe hypoxia irrespective of continuous $O_2$ monitoring, unless a backup $O_2$ delivery system is available.

**3.10 Prospective
Targeting
of End-Tidal Gases**

Prospective targeting method utilizes the principle of sequential gas delivery to control the amount and composition of gas that enters the lung and participates in gas exchange [57]. In brief, the participant first inhales gas that can partake in gas exchange, while the balance of the breath is made up of "neutral" gas. This "neutral" gas has the same composition as the participant's own alveolar gas that has equilibrated with the arterial blood, so it has no impact on gas exchange. The "neutral" gas can either be captured in a breathing reservoir during exhalation using a special sequential gas delivery circuit [57] or created by the gas blender based on participant's end-tidal values and delivered into a regular face mask [58]. The "neutral" gas feature of sequential gas delivery unlinks both PetCO$_2$ and PetO$_2$ from changes in minute ventilation as well as eliminates alveolar dead space and corresponding PetCO$_2$ to PaCO$_2$ difference, such that PetCO$_2$ actually approximates PaCO$_2$ [59]. By adjusting the $CO_2$ and $O_2$ composition of the first gas, prospective targeting method enables precise and independent control of $PetCO_2$ and $PetO_2$, as well as step, ramp, or complex (e.g., sinusoidal) changes of both gases [14]. In contrast to dynamic end-tidal forcing, prospective targeting uses gas flows that approximate resting ventilation and therefore does not require large reservoir tanks. Furthermore, in addition to emergency safety override feature that delivers 100% $O_2$ if activated, prospective targeting uses source gases that contain a minimal fraction of $O_2$ (usually 10%) to avoid accidental anoxia. All of these features have made prospective targeting the method of choice in quantitative CVR mapping in the brain [44, 60], retinal vessels [61], and spinal cord [62] of healthy volunteers, as well as in patients with cerebrovascular disease [17, 18, 63, 64], subarachnoid hemorrhage [65, 66], and concussion [23, 25, 67].

## 4    Summary and Conclusions

In summary, while many methods can be used to provide vasoactive stimulus for CVR mapping, consideration of the differences among various vasoactive stimuli in terms of *direction, magnitude, temporal profile, reproducibility,* and *safety* is critical to selecting the appropriate methods for quantitative CVR mapping. Given that one of the goals of CVR mapping is to assess cerebrovascular function in patients with pre-existing pathologies, manipulation of blood pressure using low body negative pressure or thigh cuff deflation is best avoided to prevent ischemia. Similarly, pharmacologic manipulation of blood pressure is less desirable given that it requires continuous hemodynamic monitoring with invasive intra-arterial cannulas and can induce ischemia. Manipulation of $CO_2$ may be the safest option for clinical CVR mapping, although for quantitative precision, methods that enable optimal control over magnitude, temporal pattern, and reproducibility of the $CO_2$ stimulus should be preferentially selected. While some methods such as acetazolamide administration, breath-holding, hyperventilation, and $CO_2$ inhalation are simple to implement, they do not allow precise titration of the magnitude of vasodilatory stimulus, are not easily reversible, and are challenging to reproduce within and between patients. Rebreathing, especially when combined with hyperventilation, offers a robust reproducible stimulus that spans a wide $CO_2$ range and can be employed to interrogate limits of cerebrovascular reserve. However, it only allows ramp patterns, which prevents assessment of CVR dynamics and is not suitable for use with BOLD or ASL methods. While both dynamic end-tidal forcing and prospective targeting methods enable precise titratable manipulation of $CO_2$ using both step and ramp temporal patterns, the latter method has benefits for clinical CVR mapping owing to its safety features, relative logistical simplicity, and targeting of $PaCO_2$ rather than $PetCO_2$. This likely explains why prospective targeting has recently emerged as the method of choice for quantitative CVR mapping.

## References

1. Wintermark M, Sesay M, Barbier E et al (2005) Comparative overview of brain perfusion imaging techniques. J Neuroradiol 32(5):294–314. https://doi.org/10.1016/s0150-9861(05)83159-1

2. Sobczyk O, Battisti-Charbonney A, Fierstra J et al (2014) A conceptual model for CO2-induced redistribution of cerebral blood flow with experimental confirmation using BOLD MRI. NeuroImage 92:56–68. https://doi.org/10.1016/j.neuroimage.2014.01.051

3. Ide K, Eliasziw M, Poulin MJ (2003) Relationship between middle cerebral artery blood velocity and end-tidal PCO2 in the hypocapnic-hypercapnic range in humans. J Appl Physiol Bethesda Md 1985 95(1):129–137

4. Ringelstein EB, Sievers C, Ecker S, Schneider PA, Otis SM (1988) Noninvasive assessment of CO2-induced cerebral vasomotor response in normal individuals and patients with internal carotid artery occlusions. Stroke 19(8):963–969. https://doi.org/10.1161/01.str.19.8.963

5. Bhogal AA, Siero JCW, Fisher JA et al (2014) Investigating the non-linearity of the BOLD cerebrovascular reactivity response to targeted hypo/hypercapnia at 7T. NeuroImage 98:296–305. https://doi.org/10.1016/j.neuroimage.2014.05.006

6. De Vis JB, Bhogal AA, Hendrikse J, Petersen ET, Siero JCW (2018) Effect sizes of BOLD CVR, resting-state signal fluctuations and time delay measures for the assessment of hemodynamic impairment in carotid occlusion patients. NeuroImage 179:530–539. https://doi.org/10.1016/j.neuroimage.2018.06.017

7. Battisti-Charbonney A, Fisher J, Duffin J (2011) The cerebrovascular response to carbon dioxide in humans. J Physiol 589(Pt 12):3039–3048. https://doi.org/10.1113/jphysiol.2011.206052

8. Harper AM, Glass HI (1965) Effect of alterations in the arterial carbon dioxide tension on the blood flow through the cerebral cortex at normal and low arterial blood pressures. J Neurol Neurosurg Psychiatry 28(5):449–452

9. Mutch WAC, Mandell DM, Fisher JA et al (2012) Approaches to brain stress testing: BOLD magnetic resonance imaging with computer-controlled delivery of carbon dioxide. PLoS One 7(11):e47443. https://doi.org/10.1371/journal.pone.0047443

10. Regan RE, Fisher JA, Duffin J (2014) Factors affecting the determination of cerebrovascular reactivity. Brain Behav 4(5):775–788. https://doi.org/10.1002/brb3.275

11. Duffin J, Sobczyk O, Crawley AP, Poublanc J, Mikulis DJ, Fisher JA (2015) The dynamics of cerebrovascular reactivity shown with transfer function analysis. NeuroImage 114:207–216. https://doi.org/10.1016/j.neuroimage.2015.04.029

12. Poublanc J, Crawley AP, Sobczyk O et al (2015) Measuring cerebrovascular reactivity: the dynamic response to a step hypercapnic stimulus. J Cereb Blood Flow Metab 35(11):1746–1756. https://doi.org/10.1038/jcbfm.2015.114

13. Blockley NP, Driver ID, Francis ST, Fisher JA, Gowland PA (2011) An improved method for acquiring cerebrovascular reactivity maps. Magn Reson Med 65(5):1278–1286. https://doi.org/10.1002/mrm.22719

14. Fierstra J, Sobczyk O, Battisti-Charbonney A et al (2013) Measuring cerebrovascular reactivity: what stimulus to use?: measuring cerebrovascular reactivity. J Physiol 591(23):5809–5821. https://doi.org/10.1113/jphysiol.2013.259150

15. Liu P, De Vis J, Lu H (2019) Cerebrovascular reactivity (CVR) MRI with CO2 challenge: a technical review. NeuroImage 187:104–115. https://doi.org/10.1016/j.neuroimage.2018.03.047

16. Sobczyk O, Battisti-Charbonney A, Poublanc J et al (2014) Assessing cerebrovascular reactivity abnormality by comparison to a reference atlas. J Cereb Blood Flow Metab 35:213. https://doi.org/10.1038/jcbfm.2014.184

17. Sam K, Poublanc J, Sobczyk O et al (2015) Assessing the effect of unilateral cerebral revascularisation on the vascular reactivity of the non-intervened hemisphere: a retrospective observational study. BMJ Open 5(2):

e006014.        https://doi.org/10.1136/
bmjopen-2014-006014

18. Han JS, Abou-Hamden A, Mandell DM et al
(2011) Impact of extracranial-intracranial
bypass on cerebrovascular reactivity and clinical
outcome in patients with symptomatic moyamoya        vasculopathy.        Stroke        42
(11):3047–3054.        https://doi.org/10.1161/
STROKEAHA.111.615955

19. Mandell DM, Han JS, Poublanc J et al (2011)
Quantitative measurement of cerebrovascular
reactivity by blood oxygen level-dependent
MR imaging in patients with intracranial stenosis: preoperative cerebrovascular reactivity predicts the effect of extracranial-intracranial
bypass surgery. Am J Neuroradiol 32
(4):721–727. https://doi.org/10.3174/ajnr.
A2365

20. Sobczyk O, Crawley AP, Poublanc J et al
(2016) Identifying significant changes in cerebrovascular reactivity to carbon dioxide. Am J
Neuroradiol 37(5):818–824. https://doi.org/
10.3174/ajnr.A4679

21. Peng S-L, Chen X, Li Y, Rodrigue KM, Park
DC, Lu H (2018) Age-related changes in cerebrovascular reactivity and their relationship to
cognition: a four-year longitudinal study. NeuroImage 174:257–262. https://doi.org/10.
1016/j.neuroimage.2018.03.033

22. Sam K, Conklin J, Holmes KR et al (2016)
Impaired dynamic cerebrovascular response to
hypercapnia predicts development of white
matter hyperintensities. NeuroImage Clin
11:796–801. https://doi.org/10.1016/j.nicl.
2016.05.008

23. Ellis MJ, Ryner LN, Sobczyk O et al (2016)
Neuroimaging assessment of cerebrovascular
reactivity in concussion: current concepts,
methodological considerations, and review of
the literature. Front Neurol 7:61. https://doi.
org/10.3389/fneur.2016.00061

24. Fisher JA, Venkatraghavan L, Mikulis DJ
(2018) Magnetic Resonance Imaging-Based
Cerebrovascular Reactivity and Hemodynamic
Reserve. Stroke 49(8):2011–2018. https://
doi.org/10.1161/STROKEAHA.118.
021012

25. Mutch WAC, Ellis MJ, Ryner LN et al (2016)
Brain magnetic resonance imaging CO2 stress
testing in adolescent postconcussion syndrome.    J    Neurosurg    125(3):648–660.
https://doi.org/10.3171/2015.6.JNS15972

26. Willie CK, Tzeng Y-C, Fisher JA, Ainslie PN
(2014) Integrative regulation of human brain
blood flow: integrative regulation of human
brain blood flow. J Physiol 592(5):841–859.
https://doi.org/10.1113/jphysiol.2013.
268953

27. Nishimura S, Suzuki A, Hatazawa J et al (1999)
Cerebral blood-flow responses to induced
hypotension and to CO2 inhalation in patients
with major cerebral artery occlusive disease: a
positron-emission tomography study. Neuroradiology 41(2):73–79. https://doi.org/10.
1007/s002340050709

28. Vorstrup S, Brun B, Lassen NA (1986) Evaluation of the cerebral vasodilatory capacity by the
acetazolamide test before EC-IC bypass surgery in patients with occlusion of the internal
carotid artery. Stroke 17(6):1291–1298.
https://doi.org/10.1161/01.STR.17.6.1291

29. Goswami N, Blaber AP, Hinghofer-Szalkay H,
Convertino VA (2019) Lower body negative
pressure: physiological effects, applications,
and    implementation.    Physiol    Rev    99
(1):807–851.        https://doi.org/10.1152/
physrev.00006.2018

30. Saeed NP, Horsfield MA, Panerai RB, Mistri
AK, Robinson TG (2011) Measurement of
cerebral blood flow responses to the thigh
cuff maneuver: a comparison of TCD with a
novel MRI method. J Cereb Blood Flow
Metab 31(5):1302–1310. https://doi.org/
10.1038/jcbfm.2010.225

31. Lucas SJE, Tzeng YC, Galvin SD, Thomas KN,
Ogoh S, Ainslie PN (2010) Influence of
changes in blood pressure on cerebral perfusion and oxygenation. Hypertension 55
(3):698–705.        https://doi.org/10.1161/
HYPERTENSIONAHA.109.146290

32. Kastrup A, Krüger G, Glover GH, Neumann-
Haefelin T, Moseley ME (1999) Regional
variability of cerebral blood oxygenation
response to hypercapnia. NeuroImage 10
(6):675–681.        https://doi.org/10.1006/
nimg.1999.0505

33. Silvestrini M, Vernieri F, Pasqualetti P et al
(2000) Impaired cerebral vasoreactivity and
risk of stroke in patients with asymptomatic
carotid    artery    stenosis.    JAMA    283
(16):2122–2127. https://doi.org/10.1001/
jama.283.16.2122

34. Silvestrini M, Vernieri F, Troisi E et al (1999)
Cerebrovascular reactivity in carotid artery
occlusion: possible implications for surgical
management of selected groups of patients.
Acta Neurol Scand 99(3):187–191. https://
doi.org/10.1111/j.1600-0404.1999.
tb07342.x

35. Lutfi MF (2017) The physiological basis and
clinical significance of lung volume measurements. Multidiscip Respir Med 12(1):3.
https://doi.org/10.1186/s40248-017-0084-
5

36. Markus HS, Harrison MJ (1992) Estimation of
cerebrovascular reactivity using transcranial

Doppler, including the use of breath-holding as the vasodilatory stimulus. Stroke 23 (5):668–673. https://doi.org/10.1161/01. str.23.5.668

37. Bright MG, Murphy K (2013) Reliable quantification of BOLD fMRI cerebrovascular reactivity despite poor breath-hold performance. NeuroImage 83:559–568. https://doi.org/10.1016/j.neuroimage.2013.07.007

38. Saito H, Ogasawara K, Suzuki T et al (2011) Adverse effects of intravenous acetazolamide administration for evaluation of cerebrovascular reactivity using brain perfusion single-photon emission computed tomography in patients with major cerebral artery steno-occlusive diseases. Neurol Med Chir (Tokyo) 51(7):479–483. https://doi.org/10.2176/nmc.51.479

39. Van Berkel MA, Elefritz JL (2018) Evaluating off-label uses of acetazolamide. Am J Health Syst Pharm 75(8):524–531. https://doi.org/10.2146/ajhp170279

40. Duffin J (2011) Measuring the respiratory chemoreflexes in humans. Respir Physiol Neurobiol 177(2):71–79. https://doi.org/10.1016/j.resp.2011.04.009

41. Jensen D, Wolfe LA, O'Donnell DE, Davies GAL (2005) Chemoreflex control of breathing during wakefulness in healthy men and women. J Appl Physiol 98(3):822–828. https://doi.org/10.1152/japplphysiol.01208.2003

42. Baddeley H, Brodrick PM, Taylor NJ et al (2000) Gas exchange parameters in radiotherapy patients during breathing of 2%, 3.5% and 5% carbogen gas mixtures. Br J Radiol 73 (874):1100–1104. https://doi.org/10.1259/bjr.73.874.11271904

43. Prisman E, Slessarev M, Azami T, Nayot D, Milosevic M, Fisher J (2007) Modified oxygen mask to induce target levels of hyperoxia and hypercarbia during radiotherapy: a more effective alternative to carbogen. Int J Radiat Biol 83(7):457–462. https://doi.org/10.1080/09553000701370894

44. Prisman E, Slessarev M, Han J et al (2008) Comparison of the effects of independently-controlled end-tidal $PCO_2$ and $PO_2$ on blood oxygen level-dependent (BOLD) MRI. J Magn Reson Imaging 27(1):185–191. https://doi.org/10.1002/jmri.21102

45. Ainslie PN, Duffin J (2009) Integration of cerebrovascular $CO_2$ reactivity and chemoreflex control of breathing: mechanisms of regulation, measurement, and interpretation. Am J Phys Regul Integr Comp Phys 296(5): R1473–R1495. https://doi.org/10.1152/ajpregu.91008.2008

46. McSwain SD, Hamel DS, Smith PB et al (2010) End-tidal and arterial carbon dioxide measurements correlate across all levels of physiologic dead space. Respir Care 55 (3):288–293

47. Jones NL, Robertson DG, Kane JW (1979) Difference between end-tidal and arterial $PCO_2$ in exercise. J Appl Physiol 47 (5):954–960. https://doi.org/10.1152/jappl.1979.47.5.954

48. Barr PO (1963) Pulmonary gas exchange in man as affected by prolonged gravitational stress. Acta Psychiatr Scand Suppl 207:1–46. https://doi.org/10.1111/j.1748-1716.1963.tb00082.x

49. Satoh K, Ohashi A, Kumagai M, Sato M, Kuji A, Joh S (2015) Evaluation of differences between $PaCO_2$ and $ETCO_2$ by age as measured during general anesthesia with patients in a supine position. J Anesth 2015:710537. https://doi.org/10.1155/2015/710537

50. Prause G, Hetz H, Lauda P, Pojer H, Smolle-Juettner F, Smolle J (1997) A comparison of the end-tidal-$CO_2$ documented by capnometry and the arterial $pCO_2$ in emergency patients. Resuscitation 35(2):145–148. https://doi.org/10.1016/s0300-9572(97)00043-9

51. Duffin J, Mohan RM, Vasiliou P, Stephenson R, Mahamed S (2000) A model of the chemoreflex control of breathing in humans: model parameters measurement. Respir Physiol 120(1):13–26

52. Claassen JAHR, Zhang R, Fu Q, Witkowski S, Levine BD (2007) Transcranial Doppler estimation of cerebral blood flow and cerebrovascular conductance during modified rebreathing. J Appl Physiol Bethesda Md 1985 102(3):870–877. https://doi.org/10.1152/japplphysiol.00906.2006

53. Swanson GD, Bellville JW (1975) Step changes in end-tidal CO2: methods and implications. J Appl Physiol 39(3):377–385. https://doi.org/10.1152/jappl.1975.39.3.377

54. Robbins PA, Swanson GD, Micco AJ, Schubert WP (1982) A fast gas-mixing system for breath-to-breath respiratory control studies. J Appl Physiol 52(5):1358–1362. https://doi.org/10.1152/jappl.1982.52.5.1358

55. Robbins PA, Swanson GD, Howson MG (1982) A prediction-correction scheme for forcing alveolar gases along certain time courses. J Appl Physiol 52(5):1353–1357. https://doi.org/10.1152/jappl.1982.52.5.1353

56. Wise RG, Pattinson KT, Bulte DP et al (2007) Dynamic forcing of end-tidal carbon dioxide and oxygen applied to functional magnetic resonance imaging. J Cereb Blood Flow Metab 27 (8):1521–1532. https://doi.org/10.1038/sj.jcbfm.9600465

57. Slessarev M, Han J, Mardimae A et al (2007) Prospective targeting and control of end-tidal CO2 and O2 concentrations. J Physiol 581 (Pt 3):1207–1219. https://doi.org/10.1113/jphysiol.2007.129395

58. Klein M, Fisher J (2014) Controlling arterial blood gas concentration. https://patents.google.com/patent/US10449311B2/en. Accessed 17 Dec 2020

59. Ito S, Mardimae A, Han J et al (2008) Non-invasive prospective targeting of arterial P(CO2) in subjects at rest. J Physiol 586 (15):3675–3682. https://doi.org/10.1113/jphysiol.2008.154716

60. Mark CI, Slessarev M, Ito S, Han J, Fisher JA, Pike GB (2010) Precise control of end-tidal carbon dioxide and oxygen improves BOLD and ASL cerebrovascular reactivity measures. Magn Reson Med 64(3):749–756. https://doi.org/10.1002/mrm.22405

61. Kisilevsky M, Mardimae A, Slessarev M, Han J, Fisher J, Hudson C (2008) Retinal arteriolar and middle cerebral artery responses to combined hypercarbic/hyperoxic stimuli. Invest Ophthalmol Vis Sci 49(12):5503–5509. https://doi.org/10.1167/iovs.08-1854

62. Cohen-Adad J, Gauthier CJ, Brooks JCW et al (2010) BOLD signal responses to controlled hypercapnia in human spinal cord. NeuroImage 50(3):1074–1084. https://doi.org/10.1016/j.neuroimage.2009.12.122

63. Mikulis DJ, Krolczyk G, Desal H et al (2005) Preoperative and postoperative mapping of cerebrovascular reactivity in moyamoya disease by using blood oxygen level-dependent magnetic resonance imaging. J Neurosurg 103 (2):347–355. https://doi.org/10.3171/jns.2005.103.2.0347

64. Fierstra J, Conklin J, Krings T et al (2011) Impaired peri-nidal cerebrovascular reserve in seizure patients with brain arteriovenous malformations. Brain J Neurol 134 (Pt 1):100–109. https://doi.org/10.1093/brain/awq286

65. da Costa L, Fisher J, Mikulis DJ, Tymianski M, Fierstra J (2015) Early identification of brain tissue at risk for delayed cerebral ischemia after aneurysmal subarachnoid hemorrhage. Acta Neurochir Suppl 120:105–109. https://doi.org/10.1007/978-3-319-04981-6_18

66. da Costa L, Houlden D, Rubenfeld G, Tymianski M, Fisher J, Fierstra J (2015) Impaired cerebrovascular reactivity in the early phase of subarachnoid hemorrhage in good clinical grade patients does not predict vasospasm. Acta Neurochir Suppl 120:249–253. https://doi.org/10.1007/978-3-319-04981-6_42

67. Mutch WAC, Ellis MJ, Ryner LN et al (2018) Patient-specific alterations in CO2 cerebrovascular responsiveness in acute and sub-acute sports-related concussion. Front Neurol 9:23. https://doi.org/10.3389/fneur.2018.00023

# Chapter 3

## Molecular Imaging of Cerebrovascular Reactivity

### Audrey P. Fan and Oriol Puig Calvo

### Abstract

Positron emission tomography (PET) with $[^{15}O]H_2O$ is the clinical reference standard for assessment of cerebrovascular reactivity (CVR). This molecular imaging technique measures the brain's uptake of an exogenous, radioactive tracer and fits kinetic models to these measurements to quantify perfusion at rest and after a vasoactive challenge. Single-photon emission computed tomography (SPECT) with $[^{99m}Tc]$-labelled perfusion agents provides relative (non-quantitative) regional cerebral blood flow (rCBF). This chapter describes the theory, acquisition, and modelling aspects of nuclear medicine methods to measure CVR, with focus on $[^{15}O]H_2O$ PET. The reliability of the CVR imaging biomarkers and their ability to identify hemodynamic impairment in neurovascular disorders are described. We also discuss the advantages and limitations of the approach and highlight technical advances including hybrid imaging with simultaneous PET/MRI to improve the accuracy and reduce the invasiveness of the methods.

**Key words** $[^{15}O]H_2O$, PET, SPECT, HMPAO, Cerebrovascular disease, Moyamoya, Kinetic modelling, Cerebrovascular reactivity

## 1 Introduction

The first measurements of cerebral blood flow (CBF) with molecular imaging techniques date back to 1945 when Kety and Schmidt described the nitrous oxide method to quantify CBF in non-sedated and conscious humans [1]. The method uses inhaled nitrous oxide (NO), an inert and freely diffusible tracer, and measures the tracer concentrations in the carotid artery and the jugular vein to calculate the rate of appearance and clearance of the tracer using Fick's principle. This method and its principles are still the basis of some modern techniques to quantify physiologic parameters [2] but are limited to whole-brain CBF measurements. In the 1950s, Niels Lassen and David Ingvar modified the Kety-Schmidt method to use it with radioactive inert and diffusible gases ($^{85}Kr$ or $^{133}Xe$) that were detected by an array of scintillator detectors placed around the head of the subjects [3, 4]. This permitted the mapping of regional CBF (rCBF) in humans for the first time.

Jean Chen and Jorn Fierstra (eds.), *Cerebrovascular Reactivity: Methodological Advances and Clinical Applications*, Neuromethods, vol. 175, https://doi.org/10.1007/978-1-0716-1763-2_3, © Springer Science+Business Media, LLC, part of Springer Nature 2022

Since then, several molecular imaging techniques using different tracers have been described. Ideal tracers to measure rCBF must cross the blood-brain barrier with an extraction fraction that is close to one (full extraction) and flow-independent, so their initial distribution is as close to CBF as possible. The tracer distribution must also remain unchanged in the brain long enough to enable diagnostic tomographic studies to be performed [5]. For assessment of cerebrovascular reactivity, scans are acquired at baseline and after an applied physiological challenge, so the tracer signal must decay between the two measurements.

In current clinical practice, nuclear medicine techniques to measure regional CBF typically provide only relative (non-quantitative) measurements performed with single-photon emission computerized tomography (SPECT) gamma cameras [6]. Although relative CBF measurements have been employed in the diagnostic of cerebrovascular disease [7], dementia [8] or even diagnostic of brain death [9], nowadays, they are mostly used in the presurgical localization of the epileptogenic foci in patients with drug-resistant epilepsy [10].

Fully quantitative rCBF measurements can be performed with $[^{15}O]H_2O$ in PET scanners because of the unique high sensitivity of PET detectors. PET can detect and quantify picomolar concentrations of radiotracers [11, 12] and apply appropriate tracer kinetic models to measure physiological processes. The quantitative nature of PET makes it ideally suited as a reference standard for CVR measurements. However, complete PET quantification for perfusion scans requires an arterial cannulation, on-site production of $[^{15}O]H_2O$ in a cyclotron, and a technically demanding setup that is difficult to perform clinically [13]. Despite its limitations, $[^{15}O]H_2O$ PET is currently considered the clinical reference method for rCBF measurements, and advanced techniques continue to enhance the accuracy of PET CVR scans.

## 2  Theory

### 2.1  Kinetic Model

The uptake and washout of inert, freely diffusible tracers such as NO and $[^{15}O]H_2O$ radiotracer are characterized by a one-tissue compartment model. This model was originally developed by Kety and Schmidt for gas exchange in the capillaries [1] and was later used for PET imaging of perfusion in animals [14] and humans [15]. The model relates a time course of tracer signal in the arteries (i.e., arterial input function from blood sampling) and the resulting time activity curve of signal in the tissue of interest (i.e., from dynamic PET scans in the brain). Solving the model includes fitting two time constants: the tracer delivery rate $K_1$ in mL/100 g/min, which is directly proportional to CBF, and the washout rate $k_2$ in ml/min. In theory, the fitting result numerically solves the following differential equation for $K_1$ and $k_2$:

$$\frac{\mathrm{d}C_T(t)}{\mathrm{d}t} = K_1 \cdot C_A(t) - k_2 \cdot C_T(t) \tag{1}$$

where $C_A(t)$ is the arterial input function and $C_T(t)$ is the tissue time activity curve. The analytical solution to this differential relationship is:

$$C_T(t) = K_1 \cdot C_A(t) \otimes e^{-k_2 t} = K_1 \cdot \int_0^t C_A(t) \cdot e^{-k_2(t-\tau)} \mathrm{d}\tau \tag{2}$$

where $\otimes$ denotes the convolution operator. Because the blood-brain partition coefficient $\lambda$ is equivalent to $K_1/k_2$, another common expression of the relationship can be derived:

$$C_T(t) = K_1 \cdot C_A(t) \otimes e^{-\frac{K_1 \cdot t}{\lambda}} \tag{3}$$

Several inherent assumptions underlie the one-tissue compartment model in Fig. 1. Firstly, the arterial input function (AIF) should reflect the local delivery of tracer. In practice, this AIF depends on the accuracy of radioactivity measurements in blood and its calibration to the PET scanner, as well as physiological differences in AIF shape and timing for various brain regions, during vasodilation, and in pathological conditions. Secondly, high extraction of the tracer happens instantaneously during the first pass of the tracer bolus. This assumption does not hold true for all tracers and can depend on the underlying CBF itself (i.e., change after vasodilation during a CVR test). Finally, the value for the blood-brain partition coefficient $\lambda = K_1/k_2$ is generally adopted from the literature. This value has ranged from 0.80 in white matter to 0.90–0.98 in gray matter of human subjects [16, 17] and may be

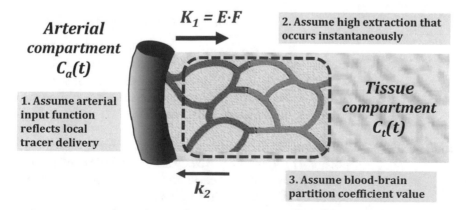

**Fig. 1** One-tissue compartment model for perfusion quantification with inert, diffusible tracers. The rate constant $K_1$ (mL/100 g/min) characterizes the delivery of tracer from the arterial blood to the tissue; and $k_2$ governs the washout rate of tracer from the tissue back to the blood compartment

abnormal in pathophysiological conditions, directly affecting the CBF quantification.

Implementation of the one-tissue compartment model is achieved using various approaches. Standard nonlinear regression compares the measured time activity curves to modelled output curves (based on the AIF) and finds the parameters $K_1$ and $k_2$ that best explain the time series data in the least-squares sense [18]. This regression can be improved with a weighted linear approach that is computationally faster and reduces estimation variability compared to the original nonlinear method [19, 20]. Alternatively, autoradiography can be used to generate a static PET scan in a defined time range $(t_1, t_2)$ with higher signal-to-noise ratio than dynamic frames. Perfusion is then measured by matching the static autoradiographic scan to a dictionary of PET tissue counts that are simulated for closely spaced values of CBF, using one of the following expressions [21]:

$$\int_{t_1}^{t_2} C_T(\tau)d\tau = \int_{t_1}^{t_2} K_1 \cdot C_A(\tau) \otimes e^{-\frac{K_1 \cdot \tau}{\lambda}} d\tau \tag{4}$$

$$\int_{t_1}^{t_2} C_T(\tau)d\tau = \int_{t_1}^{t_2} \lambda \cdot C_A(\tau) \otimes \left[k_2 \cdot e^{-k_2 \cdot \tau}\right] d\tau \tag{5}$$

Of note, Eq. 5 parameterizes the possible PET tissue count values for different washout rates $k_2$. This washout rate has also been used to noninvasively model the AIF shape to avoid the need for blood sampling in quantification of CBF and cerebrovascular reactivity [22].

A third modelling approach generates a set of basis functions using a range of physiologically plausible values for the ratio $K_1/\lambda$. The basis functions are created by convolving the AIF with the corresponding set of $e^{-\frac{K_1 \cdot t}{\lambda}}$ decay functions and take the form [23]:

$$B_i(t) = C_a(t) \otimes e^{-\frac{K_1 \cdot t}{\lambda}} \tag{6}$$

This basis function approach thus linearizes the search for $K_1$ to measure brain perfusion from the observed tissue curve $C_T(t)$. Human $[^{15}O]H_2O$ PET studies have shown that both least-squares regression and basis function approaches provide accurate estimates for CBF and perfusable tissue fraction.

## 2.2 Principles of PET Imaging

Positron emission tomography is based on the detection of pairs of photons ($\gamma$) generated from the annihilation of a positron ($e^+$) emitted from a radionuclide and an electron ($e^-$) from the tissue medium. PET radionuclides are atoms with an excess of energy in its nucleus, which makes them unstable. These radionuclides become stable after a proton ($p^+$) in the nucleus is converted into a neutron ($n$), and, from this reaction, a positron and a neutrino ($\nu$) are emitted in a process known as $\beta+$ *radioactive decay*.

$$p^+ \rightarrow n + e^+ + \nu. \tag{7}$$

The positron is then annihilated with a surrounding electron (*electron-positron annihilation*), and from this process, two photons of 511 keV are emitted in opposite directions:

$$e^+ + e^- \rightarrow \gamma + \gamma. \tag{8}$$

PET radionuclides are often combined with larger molecules in which biological distribution in the body is of interest. Such combinations are called radiotracers or radiopharmaceuticals. The most commonly used PET radiotracer is [$^{18}$F]FDG, consisting of a radionuclide of $^{18}$F at the C2 position of a glucose analogue, and it is used to study glucose metabolism [24–27].

PET scanners consist of several rings of detectors designed and optimized to detect 511 keV photons, which allow them to obtain images from any positron-emitting radionuclide. When two 511 keV photons are detected within a set time window (6–16 ns) by two detectors, they are considered coincident (originated from the same annihilation), and a *line of response* is set between the two detectors.

The information provided by all *lines of response* is later transformed into volumetric information as a result of a process called *tomographic reconstruction* that accounts and corrects for different sources of image quality degradation. *See* Fig. 2.

Clinical PET scanners nowadays are usually combined with computed tomography (CT) scanners in hybrid PET/CT scanners to combine the benefits of the metabolic or functional imaging of the PET with the better morphologic information of the CT. PET scanners can also be combined with MRI scanners in hybrid PET/MRI systems allowing simultaneous PET and MRI acquisitions, but this also introduces new challenges. PET/MRI scanners lack

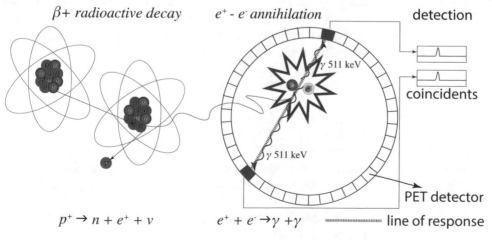

**Fig. 2** PET physics: β+ radioactive decay, annihilation, and PET detection of 511 keV photons along a line of response

CT, the current reference method for PET attenuation correction, and, due to the strong magnetic field, standard automatic blood samplers (ABBS) or automatic radiotracer injectors cannot be used in the scanner room [28]. These issues that influence quantitative $[^{15}O]H_2O$ PET studies have been addressed in recent hybrid PET/MRI studies [29, 30] and are further discussed later in this chapter.

**2.3  $[^{15}O]H_2O$ as a PET Tracer**

$^{15}O$ is a radionuclide that has a half-life of 122 s and exhibits pure $\beta+$ *radioactive decay* to $^{15}N$, which is stable.

$$^{15}O \rightarrow {}^{15}N + e^+ + \nu. \tag{9}$$

$[^{15}O]H_2O$ is synthetized by combining $^{15}O_2$ and $H_2$ either in an on-site cyclotron or in a palladium-catalyzed reaction at the bedside by an online system that also injects the radiotracer to the scanned subject. $[^{15}O]H_2O$ is an almost freely diffusible and biologically inert radiotracer that is used to study the perfusion of organs or tissues, most commonly the brain or heart, after intravenous bolus injection.

**2.4  Arterial Blood Sampling**

In order to fully quantify CBF from $[^{15}O]H_2O$ PET data, an arterial input function (AIF) is required. The AIF describes the activity concentration in the arteries as a function of time, which is one of the inputs of the kinetic model.

AIF for brain PET studies is often obtained from the catheterization of the radial artery at the wrist. Absolute contraindications for arterial cannulations are rare and include severe Raynaud's syndrome of the peripheral vasculature, subjects without proper collateral circulation of the hand and thromboangiitis obliterans. Caution should be taken in patients under anticoagulant therapy. Arterial cannulation in the wrist can be challenging, but it is a relatively safe procedure. Pain and discomfort are frequent, especially during the cannulation procedure, and complications are rare but potentially severe, in particular arterial clotting and ischemia. Additionally, the arterial catheter and posterior tubing need to be kept functional and free from clots for long periods of time in subjects that require repeated measurements (i.e., to test cerebrovascular reactivity or in activation studies), which can be challenging.

Due to the fast kinetics of the $[^{15}O]H_2O$ and its rapid decay, the AIF in $[^{15}O]H_2O$ PET studies needs to be measured continuously and quickly. Continuous sampling is best performed by automatic blood sampling systems (ABSS), which combine a pump with a previously cross-calibrated online PET detector to measure blood activity concentration. Measuring AIF with ABSS from a peripheral artery introduces delay and dispersion to the measurements that needs to be corrected.

In order to circumvent the problems associated with continuous arterial blood sampling from a peripheral artery with ABSS, several alternatives have been developed. An image-derived input function can be obtained by measuring arterial activity concentration in the big arteries within the field of view of the PET scan. Image-derived input functions from the aorta, for instance, are facilitated with recent development of total-body PET scanners that have long (2 m) axial field of view and high sensitivity to dynamically capture tracer signal in aortic vessels [31].

### 2.5 Limitations of [15O]H2O for the Assessment of CBF

The one-compartment model employed for $[^{15}O]H_2O$ PET quantification typically considers CBF as the $K_1$ of the model, assuming a complete and unrestricted water extraction through the blood-brain barrier. $[^{15}O]H_2O$, however, has a flow-dependent limitation of its extraction from the vessels in the brain, decreasing as the perfusion increases, as explained by the Crone-Renkin model [32]:

$$E = 1 - e^{-\frac{PS}{F}} \tag{10}$$

where $E$ refers to the water extraction fraction, $P$ represents the permeability of water, $S$ indicates the capillary area, and $F$ refers to the blood flow. The $PS$ product for water in the brain is assumed to be 104 mL/100 g/min [33]. In resting state ($F$ ca. 50 mL/100 g/min), the water extraction in the brain ($E$) is approximately 85%, and it decreases further as the brain blood flow increases. *See* Fig. 3.

An alternative radiotracer to $[^{15}O]H_2O$ for PET brain perfusion studies that diffuses freely through the blood-brain barrier (up to 170 mL/100 g/mL) is butanol, which can be labelled either with $^{15}O$ or with $^{11}C$ [33]. $^{11}C$ (20.38 min) has as longer half-life than $^{15}O$ (122 s) which simplifies the scanning process by reducing

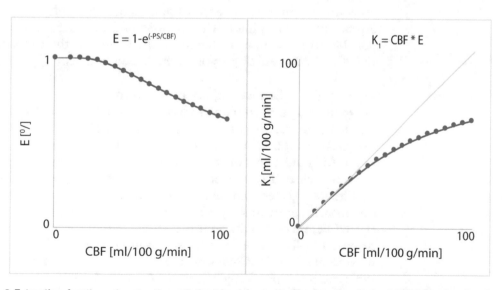

**Fig. 3** Extraction fraction of water through the blood-brain barrier as a function of CBF following the Crone-Renkin model and modelled CBF ($K_1$) as a function of real CBF when assuming full and instantaneous water extraction through the blood-brain barrier

the time pressure but results in a higher radiation dose to the scanned subject. Since more time is needed to allow for full decay of the radiotracer between measurements, fewer [11C]-butanol scans can be performed in a scanning session. [15O]-butanol, thanks to its shorter half-life, has been employed in numerous brain perfusion and activation studies with excellent results [34, 35]. The main limitation of butanol is that it needs to be synthetized for every use following a less reliable radiochemistry process compared to that of [15O]H2O.

# 3  Quantitative Cerebrovascular Reactivity Measurement with [15O]H2O PET

## 3.1  [15O]H2O PET Acquisition Considerations

### 3.1.1  Preparation

Prior to the scan, the ABSS should be calibrated to the PET detectors, and the clocks of the scanner and ABSS have to be synchronized.

Subjects that will undergo a [15O]H2O PET scan have to be able to cooperate and lay still for approximately 10 min. Absolute contraindications for [15O]H2O PET scans are pregnancy, lack of cooperation, and contraindications for arterial catheterization (*see* Subheading 2.4). Prior to the scans, subjects should avoid consuming excessive stimulants (caffeine-containing drinks or energy drinks), alcohol, and drugs that can modify CBF.

An arterial catheter has to be placed in the radial artery at the wrist (preferably of the non-dominant hand) to allow for arterial blood sampling, and a venous cannulation has to be placed in the brachial vein of the other arm for radiotracer injection.

Subjects have to be placed in the scanner headfirst with both arms out of the scanner bore to permit the manipulation of the arterial and venous lines. The ABSS should be placed as close as possible to the subject to minimize the length of the tubes and, therefore, delay and dispersion of the measurements.

### 3.1.2  Acquisition

Prior to the radiotracer injection, it is recommended to start the pump drawing arterial blood to the ABSS and the AIF recording to ensure its correct operation. Activity in the arterial blood should be sampled at a high frequency (i.e., 1 Hz) to correctly depict the peak of the AIF, and a flow of at least 8 mL/min is recommended [36].

For each PET scan, a bolus of at least 500 MBq of [15O]H2O should be injected through the venous line. The typical bolus duration is approximately 30 s. List-mode PET scans have to be started before or simultaneously with the radiotracer injection and recorded in list mode for at least 4 min and up to 10 min. The measurement of CVR requires at least two perfusion PET scans in different brain physiological states. At least 10 min has to pass between PET measurements to allow for the tracer decay over multiple half-lives.

**3.1.3  [$^{15}$O]H$_2$O PET Image Reconstruction**

[$^{15}$O]H$_2$O PET images acquired in list mode have to be reconstructed into dynamic PET studies for kinetic modelling of CBF. Short time frames are recommended in the beginning of the study in order to correctly depict the arrival of the radiotracer to the brain and can be longer at the end. While there is no consensus for what is considered an optimal reconstruction for [$^{15}$O]H$_2$O PET scans, list-mode data can be reconstructed from the injection time over 2–4 min into dynamic frames. An example of dynamic reconstruction is 18 frames of 5 s, 9 frames of 10 s, and 4 or more frames of 15 s, adjusting matrix size to achieve at least a 2 mm$^3$ voxel size, and a 2 mm Gaussian filter is recommended. The resulting image resolution with modern PET scanners should be at least 2 mm at the center of the field of view. Many PET scanners today have a three-dimensional (3D) mode to achieve higher sensitivity, because more lines of response can be acquired without the use of septa as in two-dimensional (2D) scanners. CBF quantification from 3D PET is consistent with original 2D PET measurements, only if appropriate post-processing correction for scattered photons is applied [37]. Figure 4 shows normal regional CBF studies from a healthy subject acquired in different perfusion states during a single scanning session.

[$^{15}$O]H$_2$O PET studies have to be corrected for attenuation, either with a separate transmission scan or with CT including the skull bones and sinus air passages. Other reconstruction parameters include radiotracer decay during the examination, dead time and scatter corrections, and selection of the number of iterations and subsets for commonly used OSEM (ordered subset expectation maximization) algorithms. These parameters affect the image quality and CBF quantification and should ideally be consistent between the baseline and post-vasodilation PET scans when assessing CVR. Compared to other radioisotopes, [$^{15}$O] tracers have larger positron range, which degrades the achievable spatial resolution. This resolution may be improved with newer PET detector systems that have time-of-flight capabilities, which add prior information along the detected line of response to improve localization of the tracer in the reconstructed images.

**3.1.4  Preparation of AIF**

The start time of the AIF measurement must match the start of the radiotracer injection time and corresponding PET acquisition. AIF obtained from a peripheral artery differs from the AIF in the brain arteries due to the delay and dispersion that occur while the blood travels to a peripheral artery and, more importantly, through the tubes to the ABSS. In order to minimize delay and dispersion of the AIF, it is recommended that the ABSS is located as close to the scanned subject as possible and the length of the arterial line tubes minimized. For quantitative CVR studies, AIFs are acquired for both the baseline and post-vasodilation scans, and care should be

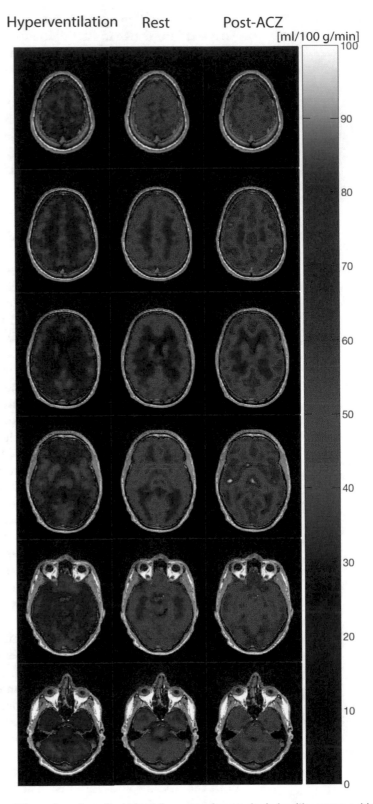

**Fig. 4** Examples of cerebral blood flow maps from a single healthy young subject obtained by [$^{15}$O]H$_2$O PET during hyperventilation and resting state and after acetazolamide (ACZ) together with T1-weighted structural MRI

taken to avoid blood clotting in the arterial line during the duration of the session.

Delay and decay corrections are usually performed during kinetic modelling using a deconvolutional procedure included in most kinetic modelling software [38]. Finally, AIF curves have to be corrected for decay of the radiotracer during the examination. A static image of the first 90 s of a $[^{15}O]H_2O$ PET scan can be used as a distribution image when arterial cannulation, and consequently absolute quantification, fails or is not available. The static PET distribution images resemble the calculated perfusion maps and can be regarded as relative CBF maps [39] but make it challenging to assess absolute CBF augmentation during a CVR vasodilation challenge.

### 3.2 Normal Values and Reproducibility of PET CVR

Using $[^{15}O]H_2O$ PET, CVR values have been reported for healthy controls as the percent augmentation in CBF due to a vasodilation challenge. Intravenous administration of acetazolamide, a carbonic anhydrase inhibitor and potent cerebral vasodilator, creates a 33% global increase in CBF as measured by PET [40]. Due to the tomographic information from PET, CVR can be assessed in individual cerebral regions, with major brain lobes including the cerebellum exhibiting robust perfusion increases [40]. There is evidence that the peak of CBF augmentation happens approximately 10 min after acetazolamide injection and stabilizes by 20 min. However, the time course of CBF changes is hard to assess with PET due to the temporal spacing between scans for tracer decay and added radiation dose of repeated $[^{15}O]H_2O$ tracer boluses.

Robust perfusion increases between 40% and 60% are also observed with $[^{15}O]H_2O$ PET during hypercapnic gas breathing challenges, e.g., 5% $CO_2$ inhalation [41]. The measured CVR depends on the modelling approach, and simplified, noninvasive approaches in particular face estimation biases compared to nonlinear regression with arterial input functions [41]. The variability of CVR assessment with PET not only depends on the imaging acquisition and reconstruction but also on the age [42, 43] and gender [44, 45] of the human subjects. Large-scale $[^{15}O]H_2O$ PET studies to investigate these demographic factors for CVR have not been performed due to the invasiveness and complexity of the procedure, especially across multiple sites.

Although day-to-day variation of baseline PET CBF scans range from 2% [46] to 9% [13] in human subjects, the data on reproducibility of CVR with PET is limited. Assessment of scan-rescan reproducibility requires at least two radiotracer injections (to measure CVR) in multiple repeat sessions (to determine reproducibility). Figure 5 shows quantitative CBF maps before and after vasodilation with 1 g of acetazolamide and the corresponding percent CVR maps in a healthy volunteer on separate sessions a

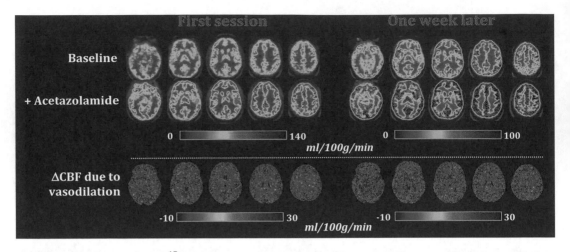

**Fig. 5** Repeat PET scans with [$^{15}$O]H$_2$O to measure CVR in a healthy control after acetazolamide vasodilation. Scans were acquired a week apart with the same dose of acetazolamide drug to highlight reproducibility. Normal CBF augmentation is observed in both scans, but with slightly different baseline CBF levels and corresponding variation in the ΔCBF response

week apart. Although the CVR distribution looks similar on both days, there are slight global differences in the CVR maps between the two sessions. It is not fully understood whether "maximal" physiological vasodilation by acetazolamide is reproducible or how it depends on drug dose and baseline CBF. Thus, CVR reproducibility by PET is an active area of research despite PET offering the reference standard method for perfusion assessment.

**3.3 PET Clinical Applications**

The main clinical application of [$^{15}$O]H$_2$O brain PET is the evaluation of ischemic cerebrovascular diseases. The term cerebrovascular disease (CVD) refers to a group of disorders that affect the blood vessels and blood supply to the brain. CVD can develop from multiple causes such as arteriosclerosis, thrombosis, embolism, aneurysm, or vascular malformations and can lead to hemorrhagic and ischemic consequences. In ischemic cerebrovascular diseases, an area of the brain is temporarily or permanently affected by a reduction of the cerebral blood flow, which causes cell damage and, eventually, cell death [47]. Cerebrovascular diseases can be acute, causing a stroke, transient and chronic as in stenosis or thrombosis.

Because it is usually not feasible to perform [$^{15}$O]H$_2$O PET scans in the acute phase, CVR assessment with PET is used to study chronic cerebrovascular conditions. In patients with chronic arterial occlusion or Moyamoya disease, measurements of cerebrovascular reactivity with [$^{15}$O]H$_2$O PET are performed to assess prognosis in patients under medical treatment [48], to select candidates for surgical treatment [49], or to assess surgical success [50].

In the clinical practice, measurements of CVR are usually performed with a pharmacological vasodilation challenge using acetazolamide. The recommended acetazolamide dose to evaluate cerebrovascular reactivity is 15 mg/kg, to be slowly administered intravenously [51]. A 30–60% CBF increase is expected in healthy subjects, and a hemodynamic response below 10% or 10 mL/100 g/min is considered clearly pathological [52]. Potential side effects of acetazolamide include slight hyperventilation, increased diuresis due to local effects on the kidneys, paresthesia (tingling or numbness) around the mouth, and slight and transient headache.

*3.3.1   Large Vessel Cerebrovascular Disease: Carotid Occlusion*

Carotid stenosis or occlusion of the internal carotid arteries is a chronic condition that may arise due to cholesterol buildup and other vascular risk factors. Different stages of hemodynamic impairment have been characterized in carotid stenosis [53]. Initially, stenosis leads to a decrease in cerebral perfusion pressure (CPP) that is adequately compensated by autoregulatory mechanisms leading to compensatory vasodilation of the arterioles. As the disease progresses and the autoregulatory mechanisms can no longer compensate for the decrease of CPP, a decrease on CBF is seen. In this situation, the cerebral metabolic rate of oxygen ($CMRO_2$) is maintained by increasing the oxygen extraction fraction (OEF), in a phenomenon known as "misery perfusion" [54]. In advanced stages of ischemic cerebrovascular disease, when CBF falls and $CMRO_2$ cannot be maintained by increasing OEF, brain ischemia occurs.

CVR imaging with acetazolamide challenge in patients with carotid stenosis can directly characterize the brain tissue pathophysiology in these patients. For instance, for brain tissue areas of decreased baseline CBF, the CVR response is impaired in some patients but is maintained in other patients, indicating severe versus moderate disease with remaining vasodilatory capability, respectively. In patients with severe CVD, areas of decreased CBF at rest can have paradoxical reduction in CBF (negative or very low CVR) after acetazolamide vasodilation, which is known as "steal phenomenon." In patients with unilateral severe stenosis or occlusion of the carotid artery, the steal phenomenon was primarily identified on the affected brain hemisphere at 5 min after the acetazolamide vasodilation [55]. This steal effect is best detected in the early phase of the acetazolamide test in patients with poor collateral (compensatory) circulation, as fewer patients show steal at 20 min after acetazolamide. Absolute quantification of CBF and CVR are required to correctly identify steal phenomenon. Figure 6 shows an example of a CVR study with [$^{15}$O]$H_2O$ PET in a patient with right internal carotid artery occlusion.

Patients presenting with steal phenomenon are the most likely to benefit from revascularization surgery. PET scans have been used to monitor improvement to CVR in carotid patients after bypass

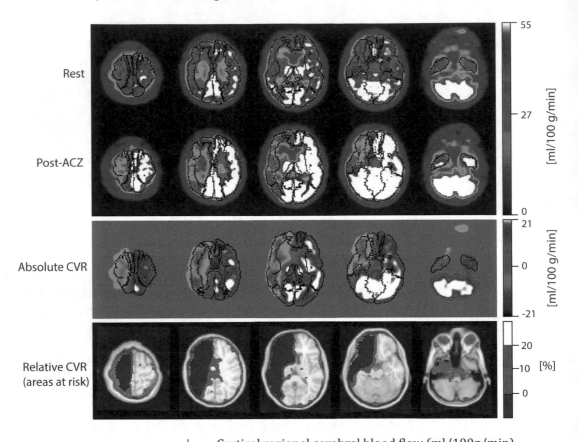

| | | Cortical regional cerebral blood flow (ml/100g/min) | | | | | |
| | | Left | | | Right | | |
| Vascular territories | Volume (ml) | Rest | Post-ACZ | CVR (%) | Rest | Post-ACZ | CVR (%) |
|---|---|---|---|---|---|---|---|
| ACA | 176 | 45 | 62 | 37 | 42 | 45 | 8 |
| MCA | 406 | 44 | 61 | 39 | 37 | 27 | -27 |
| PCA | 141 | 48 | 68 | 42 | 48 | 62 | 28 |

**Fig. 6** [¹⁵O]H₂0 PET from a 74-year-old woman with a complete right internal carotid occlusion. [¹⁵O]H₂0 PET perfusion scans at rest and after acetazolamide (rows 1 and 2) with the main vascular territories superimposed. Absolute cerebrovascular reactivity (row 3) and relative cerebrovascular reactivity thresholded into four categories (row 4) are calculated and displayed. Average cerebral blood flow in the three main vascular territories at rest and after acetazolamide and relative cerebrovascular reactivity (Change, %) are shown in the table below
ACA = anterior cerebral artery; MCA = middle cerebral artery; PCA = posterior cerebral artery; ACZ = acetazolamide

surgery, by imaging CBF at rest and after acetazolamide administration (imaging at least 5 min after vasodilation) [56]. Additionally, $[^{15}O]H_2O$ PET studies can be combined in a single scanning session with $[^{15}O]$-$O_2$ scans to obtain measurements of OEF and $CMRO_2$, two key parameters in the clinical evaluation of patients with acute and chronic cerebrovascular diseases [57].

**3.3.2  Intracranial Stenosis of Large Vessels**

Identification of CVR impairment is also clinically useful in intracranial stenosis of large cerebral arteries, including atherosclerosis of intracranial vessels and idiopathic disorders such as Moyamoya disease. Moyamoya disease in particular occurs at relatively young age [58, 59], leading to progressive narrowing of arteries at the base of the brain and extensive collateral vessel growth to compensate. Clinical manifestations include transient ischemic attacks and recurrent strokes, but few randomized clinical trials have directly investigated whether revascularization surgery benefits Moyamoya patient outcomes.

Due to the tortuous collateral patterns in Moyamoya disease, patients exhibit some of the longest arterial transit times among cerebrovascular disorders. Brain areas with long arterial transit times tend to have lower CVR in response to breathing of hypercapnic gas [60], as measured with $[^{15}O]H_2O$ PET. This inverse relationship suggests that poor collateral flow relates to impaired CVR in Moyamoya disease [60]. PET CVR imaging thus provides patient-specific information about brain pathophysiology near collateral vessels to predict the risk of future infarcts and the effect of revascularization therapy on collateral flow to brain tissues [61]. Figure 7 shows an example of MRI angiograms in a healthy control and patient with Moyamoya disease who received $[^{15}O]H_2O$ PET scans before and after acetazolamide to assess CVR. The Moyamoya patient shows graded CVR impairment that reflects the underlying level of vessel stenosis and exhibits cerebrovascular steal in the severely stenosed hemisphere, similar to other CVDs.

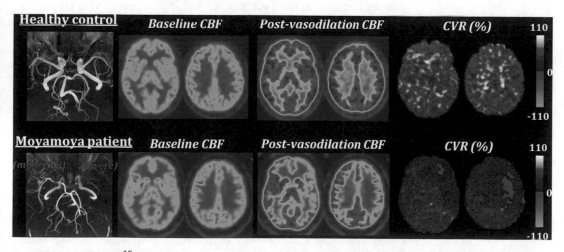

**Fig. 7** Comparison of $[^{15}O]H_2O$ PET between a healthy control and patient with bilateral Moyamoya disease, a form of idiopathic intracranial stenosis. The Moyamoya patient shows reduced CVR compared to the control on both hemispheres, with major CVR impairment in the severely stenosed vessel territories seen on MRI angiogram. Similar to carotid occlusion cases, the Moyamoya patient shows cerebrovascular steal (negative CVR) in the severely affected brain regions. (Adapted with permission from *Ishii et al., Journal of Magnetic Resonance Imaging (2020)* [85])

### 3.3.3 Cerebral Small Vessel Disease and Dementia

CVR dysfunction is also observed in neurodegenerative diseases, including Alzheimer's dementia, cognitive impairment from vascular etiologies, and cerebral small vessel disease. Reductions in CVR are more subtle in neurodegeneration and reflect underlying atherosclerosis, systemic endothelial dysfunction that affects the cerebrovascular wall lining [62], and interactions with pathological deposits of amyloid and tau [63]. Because the CVR changes expected in dementia are smaller than in CVD, earlier PET studies with hypercapnic gases likely did not have sensitivity to detect CVR differences between Alzheimer's disease patients and controls [64]. Other imaging modalities such as MRI may be preferable to evaluate CVR in neurodegenerative disorders compared to $[^{15}O]$ $H_2O$ PET, which involves complicated procedures that are less suitable for elderly populations. Acetazolamide may also produce a stronger vasodilation effect than hypercapnia and has been used with $[^{15}O]H_2O$ PET to show that anti-hypertensive drugs benefit CVR health in patients with cerebral small vessel disease [65].

## 3.4 Brain Perfusion SPECT

Brain perfusion single-photon emission computed tomography (SPECT) with $^{99m}$Tc-labelled radiotracers is a well-established, reliable, and clinically accessible method for evaluating relative regional cerebral blood flow (rCBF). Two radiotracers are most commonly used to evaluate rCBF with SPECT: technetium-99m-hexamethylpropyleneamineoxime ($[^{99m}TC]$Tc-HMPAO) and technetium-99m-ethylcysteinate dimer ($[^{99m}Tc]$Tc-ECD). Although they differ in uptake mechanism, in vitro stability, and dosimetry, in the brain, both tracers are fixed in proportion to CBF during the first minute after injection. Since there is no significant redistribution of the SPECT tracers, the tracer distribution remains unchanged for at least 2 h [5]. For this reason, if the radiotracer is administered at the time point of the desired perfusion observation, the SPECT images can be acquired later within a couple of hours. This property is especially useful in activation studies or in the detection of epilepsy onset zones during an epileptic seizure (ictal image).

The rCBF maps obtained with SPECT with $[^{99m}Tc]$Tc-HMPAO and $[^{99m}Tc]$Tc-ECD cannot be quantified in absolute units with most clinical SPECT scanners but can be statistically evaluated compared with a normal database. Currently, the main indication for SPECT studies is the presurgical localization of epileptic foci, but it has been proven useful in the assessment of cerebrovascular disease and differential diagnosis of dementias, brain trauma, or brain death.

## 3.5 Epilepsy

Epilepsy is a chronic neurological disorder caused by paroxysmal abnormal activation of a group of neurons associated with an increase of regional CBF [66]. Despite the growing availability of antiepileptic drugs, approximately 30% of patients remain refractive to medical treatment [67]. In patients with drug-resistant epilepsy, surgery is the most effective treatment, but its success depends on the correct identification of the epileptogenic zone (EZ) [68].

Brain perfusion SPECT is clinically indicated to identify the EZ. While CVR maps are not generated per se, the identification of the EZ does involve multiple SPECT scans in different brain physiological states. If patients are hospitalized in an epilepsy unit and monitored by video-electroencephalography (EEG), the radiotracer can be injected in the first seconds of an EEG-confirmed seizure to obtain an ictal brain perfusion SPECT. Ictal SPECT scans show focal hyperperfusion in the EZ [69]. Interictal images performed at least 1 week after the seizure and when no epileptogenic activity is detected on the EEG are recommended and can show normal perfusion or even hypoperfusion in the EZ [70]. In a process called SISCOM (subtraction ictal SPECT co-registered to MRI), interictal and ictal SPECT are co-registered, normalized, and subtracted to each other. The resulting difference maps are then co-registered to a structural MRI [71] to highlight hyperperfused areas during the ictal SPECT. The overall specificity and sensitivity of SISCOM to localize the EZ are approximately 94% and 85%, respectively, higher than of the ictal or interictal SPECT scans individually [72]. Figure 8 shows an example of a SISCOM study from a patient with left temporal epilepsy.

### 3.6 Cerebrovascular Disease

Brain perfusion SPECT has been validated with $[^{15}O]H_2O$ PET to identify reduced regional CVR in patients with chronic steno-occlusive CVD [73]. Additionally, several studies have demonstrated its clinical usefulness in predicting the need of surgical treatment [74] or assessing outcome after surgical treatment [75, 76] in patients with CVD. In contrast to $[^{15}O]H_2O$ PET, brain perfusion SPECT cannot be easily quantified; therefore, measurements of CVR are not as precise as with $[^{15}O]H_2O$ PET, and the detection of vascular steal can be challenging. Despite its limitations, brain perfusion SPECT remains a useful tool in patients with CVD due to its wide availability and more practical scanning method.

### 3.7 Neurodegenerative Disease

Brain perfusion is impaired early in neurodegenerative diseases before structural changes occur [77]. Different brain perfusion patterns are associated with distinct neurodegenerative syndromes, and the intensity and extension of the hypoperfusion are usually well-correlated to the clinical status [64]. For example, parietotemporal hypoperfusion is associated with Alzheimer's disease. The sensitivity for detecting perfusion alterations with SPECT is high, but the specificity in the diagnosis of neurodegenerative syndromes is relatively low as different conditions (i.e., ischemic lesions) can display patterns of hypoperfusion similar to different neurodegenerative syndromes. For instance, it is challenging to distinguish the contributions of Alzheimer's disease versus intracranial

| Ictal | Interictal | SISCOM |
|---|---|---|

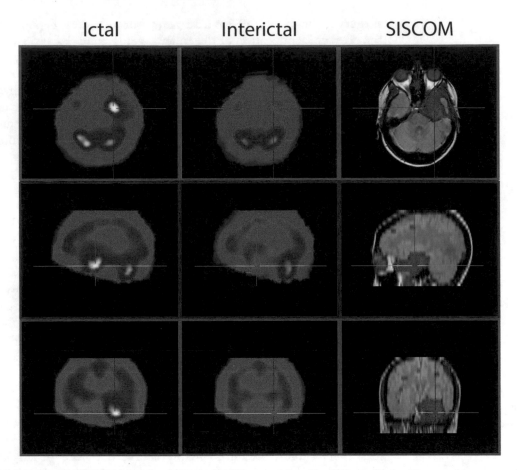

**Fig. 8** Presurgical SISCOM (subtraction ictal SPECT co-registered to MRI) study of a 39-year-old woman with drug-resistant epilepsy. Ictal and interictal SPECT are performed with [99mTc]Tc-HMPAO. Ictal images show mesial temporal focal hyperperfusion, which corresponds to an area of hypoperfusion in interictal SPECT. The differences are summarized in the SISCOM image, highlighting the epileptogenic zone (EZ)

hemorrhage due to amyloid angiopathy to brain hemodynamic failure using [99mTc]Tc-HMPAO SPECT [78]. Evaluation of neurodegenerative diseases is now usually performed with [18F]FDG PET, which measures brains' glucose metabolism, which has better diagnostic performance than brain perfusion SPECT [79].

**3.8  Traumatic Brain Injury**

Traumatic brain injury is an extremely common health problem with potentially catastrophic consequences. MRI and CT are preferred in the acute setting for their wide and rapid availability, while brain perfusion SPECT is performed in the subacute setting to assess prognosis. Brain perfusion SPECT detects hypoperfused areas due to brain injury with a higher sensitivity than structural imaging and performs better than CT or MRI at assessing long-term prognosis. A normal brain perfusion study is associated with a favorable prognosis, while the location and extent of hypoperfused lesions correlates well to clinical presentation and the severity of the sequelae to traumatic brain injury, respectively [80].

**3.9  Brain Death**

Brain death is defined as the irreversible loss of all functions of the brain, including the brainstem, and is characterized by the presence of coma, absence of brainstem reflexes, and apnea. Brain death is associated with a significant reduction in cerebral blood flow. Even though the diagnosis of brain death is based on clinical parameters, imaging diagnostics can be used to confirm the clinical suspicion, especially in patients candidate for organ donation.

In the case of brain death, dynamic planar scintigraphic studies show early arrival of activity in the internal carotid arteries, but absence of activity is seen in the anterior, middle, and posterior brain arteries and in the brain parenchyma, which can be confirmed with a SPECT ("empty light bulb" sign). In some patients, activity can be seen in the nose, known as the "hot nose" sign, or in the sagittal sinus, both of which are signs of collateral perfusion and do not rule out the diagnosis of brain death [81].

# 4  Advanced Techniques for Molecular Imaging of CVR

**4.1  PET/MRI: Reducing the Invasiveness of CVR Assessment with PET**

The advent of clinical hybrid PET/MRI scanners enables simultaneous MRI and PET exams [82, 83] that image the same brain physiological state with both modalities. This capability opens opportunities to image CVR in patients without the need for arterial blood sampling. For instance, a static PET scan lasting 90 s to 2 min after $[^{15}O]H_2O$ injection adequately represents the CBF distribution in brain tissues, but does not allow for absolute quantification to measure CVR. Instead of blood sampling, this relative PET CBF map can be scaled to quantitative units using additional information about the global blood inflow into the brain from MRI. One such MRI technique, phase-contrast MRI, provides the total blood inflow through the internal carotid and vertebral arteries based on flow velocity. Phase contrast MRI blood inflow values are consistent with global flow values from $[^{15}O]H_2O$ PET with full blood sampling [36]. By acquiring $[^{15}O]H_2O$ static PET scans and phase-contrast MRI simultaneously, the PET maps can be scaled to a measured global CBF (from phase-contrast MRI) without blood samples. If acquired both at baseline and after a physiological challenge, the PET/MRI scans can be used for noninvasive CVR assessment [84], including in cerebrovascular disease [85].

Similarly, the need for arterial blood sampling can be avoided by deriving the input function for PET kinetic modelling from PET/MRI images themselves. These image-derived input functions (IDIFs) are challenging to extract from PET alone due to the small size of arteries in the field of view (e.g., the carotid arteries) compared to the PET spatial resolution and due to signal noise in early, shorter time frames of dynamic PET series. On the other hand, PET/MRI gives added information to obtain brain

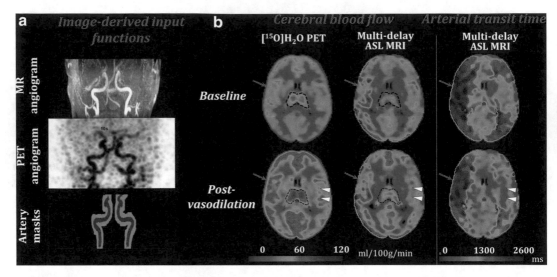

**Fig. 9** Estimation of CVR using simultaneous PET/MRI benefits from concurrent information of both modalities. (a) Image-derived input functions can be extracted from PET voxels inside the cervical arteries leading to the brain and appropriate correction using high-resolution arterial volumes seen on MRI. These IDIFs allow for quantitative CBF modelling without the need for arterial blood sampling. (b) Advanced MRI methods including arterial spin labelling can also be directly compared in both baseline and post-vasodilation states to the PET reference standard. ASL MRI requires careful attention and sequence improvements in brain areas with long arterial transit times (yellow arrows), i.e., due to cerebrovascular disease, and in regions with short arterial transit times after vasodilation (white arrowheads)

IDIFs, which would have minimal delay and dispersion relative to [$^{15}$O]H$_2$O PET uptake in cerebral tissues. For example, acquisition of simultaneous PET/MRI allows simple co-registration of high-resolution MRI angiograms that depict the carotid arteries. This MRI visualization can be used to better segment the arteries and correct for partial volume (spill-in and spill-out) effects in the carotid arteries over time [86], such that reliable PET IDIFs can be estimated. Newer PET/MRI systems also have time-of-flight PET detector capabilities, enabling high SNR in early time frames and visualization of a "PET angiogram" [87] (Fig. 9a). While still in development, the combined information of PET dynamic tracer uptake and artery structure on MRI allows repeatable, noninvasive IDIFs [88] that are suitable for quantitative perfusion mapping in CVR studies.

**4.2 Validation of Arterial Spin Labelling MRI with [$^{15}$O]H$_2$O PET**

Because of its simultaneous nature, the PET/MRI scanner is an ideal platform to also validate more advanced MRI approaches to measure CVR with the [$^{15}$O]H$_2$O PET reference standard. These advanced methods include arterial spin labelling (ASL) MRI, which use noninvasive radiofrequency labelling of blood to measure CBF, without the need for tracer or contrast injection. However, ASL MRI has not been fully validated against the PET reference standard in patients with cerebrovascular disease who may have arterial transit delays that adversely impact the ASL measurements. While ASL MRI has been compared to $[^{15}O]H_2O$ PET in various populations, these studies were mostly done in separate sessions and thus confounded by changes in cerebral physiology due to time of day, caffeine intake, and diet, among other factors.

Using simultaneous PET/MRI, recent investigators have directly compared CBF acquired at the same time with ASL MRI and $[^{15}O]H_2O$ PET. These validations provide critical data to optimize ASL MRI strategies for challenging cerebrovascular patient cases [89], so that noninvasive MRI methods can be clinically applied. Importantly, the accuracy of CBF from each imaging modality is affected differently when brain physiology changes, so it is vital for ASL MRI and PET to be compared in multiple physiological states for CVR assessment. Recent studies have shown good correspondence between ASL and PET CBF maps both at baseline and after vasodilation with hypercapnia [90] and acetazolamide [91]. Residual CVR differences between ASL and MRI can occur in the post-vasodilation state due to faster arterial arrival times (Fig. 9b) and motivate further ASL MRI improvements in the future.

**4.3 Machine Learning Prediction of CVR**

Machine learning algorithms are gaining broader use and attention in medical imaging because they perform well in image enhancement and clinical prediction tasks. Although exploratory in nature, CVR measurements from molecular imaging can be improved by using machine learning to increase the SNR of ("denoise") PET images that were acquired with very low radiotracer doses [92, 93]. This reduces the overall radiation exposure to the patient, especially as multiple scans are required in CVR studies, while still providing clinically readable PET images comparable to those obtained with a full tracer dose.

CVR can also be predicted by machine learning algorithms based on the underlying PET images as inputs. This prediction task is well-suited to a subset of machine learning approaches called deep learning, which use multiple layers in a convolutional network structure to extract hidden features from input images. This deep learning paradigm can synthesize complex features of PET/MRI data, leveraging both image modalities to make predictions of CVR. An initial study showed that by using PET or PET/MRI with $[^{15}O]H_2O$ tracer, CVR maps could be predicted from baseline images of CBF with deep learning, avoiding the need to administer a vasodilatory stress test altogether [94]. Improvements to scanner hardware (increased PET sensitivity and multimodal scans) and the ability to synergize this molecular imaging data thus continue to advance capabilities for reliable, noninvasive CVR measurements with PET.

## References

1. Kety SS, Schmidt CF (1948) The nitrous oxide method for the quantitative determination of cerebral blood flow in man: theory, procedure and normal values. J Clin Invest 27:476–483
2. Xu F, Ge Y, Lu H (2009) Noninvasive quantification of whole-brain cerebral metabolic rate of oxygen (CMRO2) by MRI. Magn Reson Med 62:141–148
3. Ingvar DH, Lassen NA (1961) Quantitative determination of cerebral blood flow in man. Lancet 2:806–807
4. Lassen NA, Ingvar DH (1972) Radioisotopic assessment of regional cerebral blood flow. Prog Nucl Med 1:376–409
5. Kapucu OL, Nobili F, Varrone A et al (2009) EANM procedure guideline for brain perfusion SPECT using 99mTc-labelled radiopharmaceuticals, version 2. Eur J Nucl Med Mol Imaging 36:2093–2102
6. Catafau AM (2001) Brain SPECT in clinical practice. Part I: Perfusion. J Nucl Med 42:259–271
7. Sugawara Y, Kikuchi T, Ueda T et al (2002) Usefulness of brain SPECT to evaluate brain tolerance and hemodynamic changes during temporary balloon occlusion test and after permanent carotid occlusion. J Nucl Med 43:1616–1623
8. Matsuda H (2007) Role of neuroimaging in Alzheimer's disease, with emphasis on brain perfusion SPECT. J Nucl Med 48:1289–1300
9. Spieth ME, Devadas GC, Gauger BS (2004) Procedure guideline for brain death scintigraphy. J Nucl Med 45:922. author reply 922
10. Kazemi NJ, Worrell GA, Stead SM et al (2010) Ictal SPECT statistical parametric mapping in temporal lobe epilepsy surgery. Neurology 74:70–76
11. Hoffman EJ, Guerrero TM, Germano G et al (1989) PET system calibrations and corrections for quantitative and spatially accurate images. IEEE Trans Nucl Sci 36:1108–1112
12. Boellaard R (2009) Standards for PET image acquisition and quantitative data analysis. J Nucl Med 50(Suppl 1):11S–20S
13. Bremmer JP, van Berckel BN, Persoon S et al (2011) Day-to-day test-retest variability of CBF, CMRO2, and OEF measurements using dynamic 15O PET studies. Mol Imaging Biol 13:759–768
14. Landau WM, Freygang WH, Roland LP et al (1955) The local circulation of the living brain; values in the unanesthetized and anesthetized cat. Trans Am Neurol Assoc 80:125–129
15. Raichle ME, Martin WR, Herscovitch P et al (1983) Brain blood flow measured with intravenous H2(15)O. II. Implementation and validation. J Nucl Med 24:790–798
16. Larsson HB, Courivaud F, Rostrup E et al (2009) Measurement of brain perfusion, blood volume, and blood-brain barrier permeability, using dynamic contrast-enhanced T(1)-weighted MRI at 3 tesla. Magn Reson Med 62:1270–1281
17. Herscovitch P, Raichle ME (1985) What is the correct value for the brain--blood partition coefficient for water? J Cereb Blood Flow Metab 5:65–69
18. Huang SC, Mahoney DK, Phelps ME (1987) Quantitation in positron emission tomography: 8. Effects of nonlinear parameter estimation on functional images. J Comput Assist Tomogr 11:314–325
19. Zhou Y, Huang SC, Bergsneider M (2001) Linear ridge regression with spatial constraint for generation of parametric images in dynamic positron emission tomography studies. IEEE Trans Nucl Sci 48:125–130
20. Ho D, Feng D (1999) Rapid algorithms for the construction of cerebral blood flow and oxygen utilization images with oxygen-15 and dynamic positron emission tomography. Comput Methods Prog Biomed 58:99–117
21. Kanno I, Iida H, Miura S et al (1987) A system for cerebral blood flow measurement using an H215O autoradiographic method and positron emission tomography. J Cereb Blood Flow Metab 7:143–153
22. Treyer V, Jobin M, Burger C et al (2003) Quantitative cerebral H2(15)O perfusion PET without arterial blood sampling, a method based on washout rate. Eur J Nucl Med Mol Imaging 30:572–580
23. Boellaard R, Knaapen P, Rijbroek A et al (2005) Evaluation of basis function and linear least squares methods for generating parametric blood flow images using 15O-water and positron emission tomography. Mol Imaging Biol 7:273–285
24. Al-Ibraheem A, Buck A, Krause BJ et al (2009) Clinical applications of FDG PET and

PET/CT in head and neck cancer. J Oncol 2009:208725

25. Shivamurthy VK, Tahari AK, Marcus C et al (2015) Brain FDG PET and the diagnosis of dementia. AJR Am J Roentgenol 204: W76–W85

26. Bruzzi JF, Munden RF, Truong MT et al (2007) PET/CT of esophageal cancer: its role in clinical management. Radiographics 27:1635–1652

27. Subhas N, Patel PV, Pannu HK et al (2005) Imaging of pelvic malignancies with in-line FDG PET-CT: case examples and common pitfalls of FDG PET. Radiographics 25:1031–1043

28. Andersen FL, Ladefoged CN, Beyer T et al (2014) Combined PET/MR imaging in neurology: MR-based attenuation correction implies a strong spatial bias when ignoring bone. NeuroImage 84:206–216

29. Ladefoged CN, Law I, Anazodo U et al (2016) A multi-centre evaluation of eleven clinically feasible brain PET/MRI attenuation correction techniques using a large cohort of patients. NeuroImage 147:346–359

30. Zhang K, Herzog H, Mauler J et al (2014) Comparison of cerebral blood flow acquired by simultaneous [15O]water positron emission tomography and arterial spin labeling magnetic resonance imaging. J Cereb Blood Flow Metab 34:1373–1380

31. Zhang X, Xie Z, Berg E et al (2020) Total-body dynamic reconstruction and parametric imaging on the uEXPLORER. J Nucl Med 61:285–291

32. Lorthois S, Duru P, Billanou I et al (2014) Kinetic modeling in the context of cerebral blood flow quantification by H2(15)O positron emission tomography: the meaning of the permeability coefficient in Renkin-Crones model revisited at capillary scale. J Theor Biol 353:157–169

33. Herscovitch P, Raichle ME, Kilbourn MR et al (1987) Positron emission tomographic measurement of cerebral blood flow and permeability-surface area product of water using [15O]water and [11C]butanol. J Cereb Blood Flow Metab 7:527–542

34. Herzog H, Seitz RJ, Tellmann L et al (1996) Quantitation of regional cerebral blood flow with 15O-butanol and positron emission tomography in humans. J Cereb Blood Flow Metab 16:645–649

35. Wunderlich G, Knorr U, Stephan KM et al (1997) Dynamic scanning of 15O-butanol with positron emission tomography can identify regional cerebral activations. Hum Brain Mapp 5:364–378

36. Puig O, Vestergaard MB, Lindberg U et al (2019) Phase contrast mapping MRI measurements of global cerebral blood flow across different perfusion states - a direct comparison with (15)O-H2O positron emission tomography using a hybrid PET/MR system. J Cereb Blood Flow Metab 39:2368–2378

37. Ibaraki M, Miura S, Shimosegawa E et al (2008) Quantification of cerebral blood flow and oxygen metabolism with 3-dimensional PET and 15O: validation by comparison with 2-dimensional PET. J Nucl Med 49:50–59

38. Meyer E (1989) Simultaneous correction for tracer arrival delay and dispersion in CBF measurements by the H215O autoradiographic method and dynamic PET. J Nucl Med 30:1069–1078

39. Kanno I, Miura S, Murakami M (1991) Optimal scan time of oxygen- 15-labeled water cerebral blood flow. J Nucl Med 32:1931–1934

40. Okazawa H, Yamauchi H, Sugimoto K et al (2001) Effects of acetazolamide on cerebral blood flow, blood volume, and oxygen metabolism: a positron emission tomography study with healthy volunteers. J Cereb Blood Flow Metab 21:1472–1479

41. Koopman T, Yaqub M, Heijtel DF et al (2019) Semi-quantitative cerebral blood flow parameters derived from non-invasive [(15)O] H2O PET studies. J Cereb Blood Flow Metab 39:163–172

42. Thomas BP, Liu P, Park DC et al (2014) Cerebrovascular reactivity in the brain white matter: magnitude, temporal characteristics, and age effects. J Cereb Blood Flow Metab 34:242–247

43. Peng SL, Chen X, Li Y et al (2018) Age-related changes in cerebrovascular reactivity and their relationship to cognition: a four-year longitudinal study. NeuroImage 174:257–262

44. Deegan BM, Sorond FA, Lipsitz LA et al (2009) Gender related differences in cerebral autoregulation in older healthy subjects. Annu Int Conf IEEE Eng Med Biol Soc 2009:2859–2862

45. Olah L, Valikovics A, Bereczki D et al (2000) Gender-related differences in acetazolamide-induced cerebral vasodilatory response: a transcranial Doppler study. J Neuroimaging 10:151–156

46. Coles JP, Fryer TD, Bradley PG et al (2006) Intersubject variability and reproducibility of 15O PET studies. J Cereb Blood Flow Metab 26:48–57

47. Van Lieshout JJ, Wieling W, Karemaker JM et al (2003) Syncope, cerebral perfusion, and oxygenation. J Appl Physiol (1985) 94:833–848

48. Kuroda S, Houkin K, Kamiyama H et al (2001) Long-term prognosis of medically treated patients with internal carotid or middle cerebral artery occlusion: can acetazolamide test predict it? Stroke 32:2110–2115

49. Powers WJ, Clarke WR et al (2011) Extracranial-Intracranial Bypass Surgery for Stroke Prevention in Hemodynamic Cerebral Ischemia The Carotid Occlusion Surgery Study Randomized Trial. JAMA 306:1983–1992

50. Kuhn FP, Warnock G, Schweingruber T et al (2015) Quantitative H2[(15)O]-PET in pediatric moyamoya disease: evaluating perfusion before and after cerebral revascularization. J Stroke Cerebrovasc Dis 24:965–971

51. Vagal AS, Leach JL, Fernandez-Ulloa M et al (2009) The acetazolamide challenge: techniques and applications in the evaluation of chronic cerebral ischemia. AJNR Am J Neuroradiol 30:876–884

52. Hosoda K, Kawaguchi T, Shibata Y et al (2001) Cerebral vasoreactivity and internal carotid artery flow help to identify patients at risk for hyperperfusion after carotid endarterectomy. Stroke 32:1567–1573

53. Derdeyn CP, Videen TO, Yundt KD et al (2002) Variability of cerebral blood volume and oxygen extraction: stages of cerebral haemodynamic impairment revisited. Brain 125:595–607

54. Yamauchi H, Fukuyama H, Nagahama Y et al (1996) Evidence of misery perfusion and risk for recurrent stroke in major cerebral arterial occlusive diseases from PET. J Neurol Neurosurg Psychiatry 61:18–25

55. Kuwabara Y, Ichiya Y, Sasaki M et al (1995) Time dependency of the acetazolamide effect on cerebral hemodynamics in patients with chronic occlusive cerebral arteries. Early steal phenomenon demonstrated by [15O]H2O positron emission tomography. Stroke 26:1825–1829

56. Kuwabara Y, Ichiya Y, Sasaki M et al (1998) PET evaluation of cerebral hemodynamics in occlusive cerebrovascular disease pre- and post-surgery. J Nucl Med 39:760–765

57. Grubb RL, Derdeyn CP, Fritsch SM et al (1998) Importance of hemodynamic factors in the prognosis of symptomatic carotid occlusion. J Am Med Assoc 280:1055–1060

58. Bruno A, Adams HP, Biller J et al (1988) Cerebral infarction due to moyamoya disease in young adults. Stroke 19:826–833

59. Kim SK, Seol HJ, Cho BK et al (2004) Moyamoya disease among young patients: its aggressive clinical course and the role of active surgical treatment. Neurosurgery 54:840–844. discussion 844-846

60. Kuwabara Y, Ichiya Y, Sasaki M et al (1997) Response to hypercapnia in moyamoya disease. Cerebrovascular response to hypercapnia in pediatric and adult patients with moyamoya disease. Stroke 28:701–707

61. Kuroda S, Kashiwazaki D, Hirata K et al (2014) Effects of surgical revascularization on cerebral oxygen metabolism in patients with Moyamoya disease: an 15O-gas positron emission tomographic study. Stroke 45:2717–2721

62. Glodzik L, Randall C, Rusinek H et al (2013) Cerebrovascular reactivity to carbon dioxide in Alzheimer's disease. J Alzheimers Dis 35:427–440

63. Stefani A, Sancesario G, Pierantozzi M et al (2009) CSF biomarkers, impairment of cerebral hemodynamics and degree of cognitive decline in Alzheimer's and mixed dementia. J Neurol Sci 283:109–115

64. Jagust WJ, Eberling JL, Reed BR et al (1997) Clinical studies of cerebral blood flow in Alzheimer's disease. Ann N Y Acad Sci 826:254–262

65. Kimura Y, Kitagawa K, Oku N et al (2010) Blood pressure lowering with valsartan is associated with maintenance of cerebral blood flow and cerebral perfusion reserve in hypertensive patients with cerebral small vessel disease. J Stroke Cerebrovasc Dis 19:85–91

66. Penfield W (1933) The evidence for a cerebral vascular mechanism in epilepsy. Ann Intern Med 7:303–310

67. Kwan P, Brodie MJ (2000) Early identification of refractory epilepsy. N Engl J Med 342:314–319

68. Dwivedi R, Ramanujam B, Chandra PS et al (2017) Surgery for drug-resistant epilepsy in children. N Engl J Med 377:1639–1647

69. Knowlton RC, Lawn ND, Mountz JM et al (2004) Ictal SPECT analysis in epilepsy. Neurology 63:10–15

70. Grunwald F, Menzel C, Pavics L et al (1994) Ictal and interictal brain SPECT imaging in epilepsy using technetium-99m-ECD. J Nucl Med 35:1896–1901

71. Chen T, Guo L (2016) The role of SISCOM in preoperative evaluation for patients with epilepsy surgery: a meta-analysis. Seizure 41:43–50

72. Newey CR, Wong C, Irene Wang Z et al (2013) Optimizing SPECT SISCOM analysis

to localize seizure-onset zone by using varying z scores. Epilepsia 54:793–800

73. Ogasawara K, Ito H, Sasoh M et al (2003) Quantitative measurement of regional cerebrovascular reactivity to acetazolamide using 123I-N-isopropyl-p-iodoamphetamine autoradiography with SPECT: validation study using H2 15O with PET. J Nucl Med 44:520–525

74. Jae Seung K, Dae Hyuk M, Geun Eun K et al (2000) Acetazolamide stress brain-perfusion SPECT predicts the need for carotid shunting during carotid endarterectomy. J Nucl Med 41:1836–1841

75. Lee HY, Jin CP, Dong SL et al (2004) Efficacy assessment of cerebral arterial bypass surgery using statistical parametric mapping and probabilistic brain atlas on basal/acetazolamide brain perfusion SPECT. J Nucl Med 45:202–206

76. Cikrit DF, Dalsing MC, Harting PS et al (1997) Cerebral vascular reactivity assessed with acetazolamide single photon emission computer tomography scans before and after carotid endarterectomy. Am J Surg 174:193–197

77. Morinaga A, Ono K, Ikeda T et al (2010) A comparison of the diagnostic sensitivity of MRI, CBF-SPECT, FDG-PET and cerebrospinal fluid biomarkers for detecting Alzheimer's disease in a memory clinic. Dement Geriatr Cogn Disord 30:285–292

78. Mehdorn HM, Gerhard L, Muller SP et al (1992) Clinical and cerebral blood flow studies in patients with intracranial hemorrhage and amyloid angiopathy typical of Alzheimer's disease. Neurosurg Rev 15:111–116

79. O'Brien JT, Firbank MJ, Davison C et al (2014) 18F-FDG PET and perfusion SPECT in the diagnosis of Alzheimer and Lewy body dementias. J Nucl Med 55:1959–1965

80. Raji CA, Tarzwell R, Pavel D et al (2014) Clinical utility of SPECT neuroimaging in the diagnosis and treatment of traumatic brain injury: a systematic review. PLoS One 9:e91088

81. Zuckier LS, Kolano J (2008) Radionuclide studies in the determination of brain death: criteria, concepts, and controversies. Semin Nucl Med 38:262–273

82. Werner P, Barthel H, Drzezga A et al (2015) Current status and future role of brain PET/MRI in clinical and research settings. Eur J Nucl Med Mol Imaging 42:512–526

83. Catana C, Drzezga A, Heiss WD et al (2012) PET/MRI for neurologic applications. J Nucl Med 53:1916–1925

84. Ssali T, Anazodo UC, Thiessen JD et al (2018) A noninvasive method for quantifying cerebral blood flow by hybrid PET/MRI. J Nucl Med 59:1329–1334

85. Ishii Y, Thamm T, Guo J et al (2020) Simultaneous phase-contrast MRI and PET for noninvasive quantification of cerebral blood flow and reactivity in healthy subjects and patients with cerebrovascular disease. J Magn Reson Imaging 51:183–194

86. Su Y, Arbelaez AM, Benzinger TL et al (2013) Noninvasive estimation of the arterial input function in positron emission tomography imaging of cerebral blood flow. J Cereb Blood Flow Metab 33:115–121

87. Khalighi MM, Deller TW, Fan AP et al (2018) Image-derived input function estimation on a TOF-enabled PET/MR for cerebral blood flow mapping. J Cereb Blood Flow Metab 38:126–135

88. Andersen JB, Lindberg U, Olesen OV et al (2019) Hybrid PET/MRI imaging in healthy unsedated newborn infants with quantitative rCBF measurements using (15)O-water PET. J Cereb Blood Flow Metab 39:782–793

89. Fan AP, Guo J, Khalighi MM et al (2017) Long-delay arterial spin labeling provides more accurate cerebral blood flow measurements in moyamoya patients: a simultaneous positron emission tomography/MRI study. Stroke 48:2441–2449

90. Puig O, Henriksen OM, Vestergaard MB et al (2020) Comparison of simultaneous arterial spin labeling MRI and (15)O-H2O PET measurements of regional cerebral blood flow in rest and altered perfusion states. J Cereb Blood Flow Metab 40:1621–1633

91. Okazawa H, Higashino Y, Tsujikawa T et al (2018) Noninvasive method for measurement of cerebral blood flow using O-15 water PET/MRI with ASL correlation. Eur J Radiol 105:102–109

92. Chen KT, Gong E, de Carvalho Macruz FB et al (2020) Ultra-Low-Dose (18)F-Florbetaben Amyloid PET Imaging Using Deep Learning with Multi-Contrast MRI Inputs. Radiology 296:E195

93. Kaplan S, Zhu YM (2019) Full-dose PET image estimation from low-dose PET image using deep learning: a pilot study. J Digit Imaging 32:773–778

94. Chen DYT, Ishii Y, Fan AP et al (2020) Predicting PET cerebrovascular reserve with deep learning by using baseline MRI: a pilot investigation of a drug-free brain stress test. Radiology 296:627–637

# Chapter 4

# Measurement of Cerebrovascular Reactivity Using Transcranial Doppler

## Leodante da Costa and Martin Chapman

## Abstract

The Doppler effect (or Doppler shift), which is the basis for the transcranial Doppler (TCD) technique, was described by Christian Andreas Doppler, an Austrian physicist, at a meeting of the Natural Sciences Section of the Royal Bohemian Society in Prague on 25 May 1842. The principle was presented in the paper *Über das farbige Licht der Doppelsterne* (Eden, Ultrasound Med Biol 16:831–832, 1990) (*Concerning the colored light of the double stars*) and initially applied to astronomy. In 1965, M. Miyazaki and K. Kato (Jpn Circ J 29:375–382, 1965) described the use of ultrasonic Doppler technique in the evaluation of blood flow and hemodynamics. However, ultrasound technology at the time was not able to penetrate the skull, and therefore cerebral blood flow (CBF) could not be assessed directly. With the development of the low frequency pulsed Doppler technique (2 MHz), able to penetrate the calvarium in most skulls, Aaslid et al. (J Neurosurg 57:769–774, 1982) were able to use TCD to measure blood flow velocity in the intracranial arteries for the first time. Since then, TCD has been used in medical practice as a technique to measure CBF and later cerebrovascular reactivity (CVR) in many conditions, including ischemic stroke, TBI, subarachnoid hemorrhage and vasospasm, and brain death, both in clinical practice and as a research tool in the search of better understanding physiologic responses of the intracranial circulation to healthy and pathologic stimuli. The objective of this chapter is to provide a brief summary of the use of TCD in the evaluation of cerebrovascular reactivity, with a brief overview of its use in TBI and aneurysmal subarachnoid hemorrhage.

**Keywords** Brain aneurysm, Cerebrovascular reactivity, Subarachnoid hemorrhage, Transcranial Doppler, Traumatic brain injury, Ultrasound

## 1    Introduction

Transcranial Doppler (TCD) is based on the Doppler effect, or Doppler shift, described by Christian Doppler in 1842 [1]. He noticed that the frequency of sound waves changed in relation to an observer who's moving relative to the wave source. During a vascular examination using a Doppler probe, the ultrasonic waves generated by the probe will be deflected by the moving red blood cells within the blood vessels. The difference in the frequency between the emitted and reflected waves (the Doppler shift) is

Jean Chen and Jorn Fierstra (eds.), *Cerebrovascular Reactivity: Methodological Advances and Clinical Applications*, Neuromethods, vol. 175, https://doi.org/10.1007/978-1-0716-1763-2_4, © Springer Science+Business Media, LLC, part of Springer Nature 2022

directly proportional to the velocity of the moving cells. A major breakthrough that allowed the use of this technology to evaluate intracranial hemodynamics was the development of the low-frequency pulsed Doppler technique (2 MHz) able to penetrate the calvarium (in most but not all skulls) by Aaslid et al. [3]. With this, TCD, being a relatively inexpensive, portable, and fast way to obtain real-time physiological information on CBF, became a popular way of evaluating cerebral hemodynamics.

Transcranial Doppler examination is a real-time, dynamic study that can be influenced by many technical and physiologic factors, including the angle of insonation, age, hematocrit, carbon dioxide ($CO_2$), blood pressure, and activity. The response of the intracranial blood vessels to some of these variables can be used to investigate the condition and health of the brain circulation. In this chapter, we will focus on the use of $CO_2$ as the vasoactive stimulus. The influence of $CO_2$ on the intracranial vasculature is long known. The vasodilatory effects of increasing $CO_2$ levels were described in the 1920s by Bronk and Gesell [4], and 20 years later, Kety and Schmidt [5] provided conclusive evidence of $CO_2$ effects on cerebral blood flow.

Carbon dioxide manipulation has been our method of choice to evaluate CVR in many conditions. Note that cerebral autoregulation (pressure-dependent changes in intracranial vessel diameter) and neurovascular coupling, the other major mechanisms of control in cerebral blood flow, are not discussed in detail. The following sections describe the TCD examination, the mechanisms of $CO_2$ control and its influence on the cerebrovascular tree, and the available data on the clinical uses of TCD as a tool to investigate cerebrovascular reactivity.

## 2    Transcranial Doppler Examination

Transcranial Doppler measures the blood flow velocity within a given vessel. Although it can be used to access many large vessels at the skull base, the blood vessel most widely used is the MCA because of its favorable orientation and accessibility to TCD [6].

Sound waves that leave a source (Doppler probe in this case) expand in spherical waves centered on the source. The Doppler effect describes the apparent difference between the frequency at which these waves leave a source and that at which they reach an observer. This difference is caused by relative motion of the observer and the wave source. For a moving observer (or source), the received frequency is higher (compared to the emitted frequency) during the approach, it is identical to the emitted frequency at the instant of passing by, and it is lower than the emitting frequency as the observer moves away from the source. When the distance between the observer and the source is decreasing while they approach each other, each successive wave is emitted from a closer position than the previous wave, taking less time to reach the observer than the previous one. The decreased time between successive waves leads to an increase in the frequency. The opposite is true when the distance between observer and source is increasing. Transcranial Doppler uses the Doppler effect to measure CBF velocity. While it can be used as a surrogate of CBF, it is not a direct measure of CBF.

The TCD examination is performed using a low-frequency probe (2 MHz). These probes generate a wave frequency that is able to penetrate the skull in most patients. Even with the use of low-frequency probes, insonation of the intracranial arteries is only possible on patients with areas of thinning in the skull, the so-called bone windows. Four major bone windows can be used: transtemporal, transorbital, submandibular, and suboccipital [7]. Each of these windows grants access to specific areas of the intracranial circulation, and although all four windows are used for a complete TCD examination, in routine clinical management, i.e., monitoring for vasospasm after aneurysmal subarachnoid hemorrhage (aSAH), the temporal and suboccipital windows are the ones used routinely.

The temporal window allows access to the internal carotid artery (ICA), ICA bifurcation, middle cerebral artery (MCA) and proximal anterior cerebral artery (ACA), and posterior cerebral artery (PCA), being therefore very useful to monitor physiologic and pathologic responses of the intracranial circulation. Unfortunately, these windows are absent in 8% to up to 25% of patients [8].

Cerebrovascular reactivity is often assessed using the temporal window. Through this window, vessels in the anterior circulation and the PCA can be identified. The probe is placed over the temporal region, just above the zygomatic arch and aiming superiorly and anteriorly. With knowledge of the vascular anatomy at the circle of Willis, changes in probe orientation and ultrasound depth allow for mapping of the vessel of interest. The first vessel to be identified is usually the ipsilateral MCA, which can then be used as a reference to identify the ICA bifurcation and the ACA. By changing the depth of insonation, the MCA can be "followed" medially to the bifurcation. A combination of different depths, angles, and flow direction allows for identification of the ACA and ICA, as well as PCAs [9]. In cases where distinguishing between the anterior and posterior circulation is difficult, response to careful carotid compression can be used. M. Hennerici et al. provide a succinct and very good review of TCD examination and normal blood flow values [9].

Testing CVR with TCD involves first identifying the intracranial blood vessel of interest, usually the MCA, and establishing baseline measurement. The desired stimulus is then applied to measure the change in the baseline. Tests were done in a standard hospital bed, with patients resting without moving or talking, unless the stimuli being tested involve a motor task. Ideally, bilateral TCD probes are used. The use of a head frame to fix the probes in place facilitates the test and minimizes the risk of probe movement. Once the vessel of interest is identified, the proper transducer position and angulation are checked, and probes are fixed on the head frame to prevent motion. The MCA can be identified using established TCD criteria: depth of insonation from 30 to 60 mm, flow direction toward the transducer, mean flow velocities from 46 to 86 cm/s, and anatomical relationships with the ICA bifurcation.

Doppler measurements provide unidimensional results, usually in cm/s. Because CBF is a three-dimensional measurement of volume of blood per unit of tissue per unit of time (mL/100 g/min), MCA flow velocity (MCAv) and CBF are two intrinsically different values. Therefore, the correlation of absolute MCAv values and CBF measurements is poor, and absolute values of MCAv cannot be used as a surrogate of CBF. However, it has been demonstrated that the diameter of the large vessels at the circle of Willis remains relatively constant during changes in blood pressure and $CO_2$ levels. If the caliber of the vessel where measurements are being made is constant, any changes in MCAv must reflect changes in CBF. These changes happen due to changes in cerebrovascular resistance in the distal microcirculation and the consequent decrease or increase of flow caused by constriction or dilatation of the microcirculation in response to the applied stimulus. This relationship means that *changes* in MCAv can be used to estimate *changes* in CBF [10]. This correlation seems to be stronger at CBF levels away from the extremes of the autoregulation curve [6].

## 3 Cerebral Blood Flow, Cerebral Blood Flow Velocity, and CBF Manipulation

The human brain is a relatively small organ but consumes up to 20% of all oxygen delivered throughout the body in each cardiac cycle. This high metabolic demand requires a constant, stable delivery of oxygen and glucose to generate energy and maintain function [11]. Many mechanisms are involved in controlling CBF and the associated supply of oxygen and nutrients to the brain. These mechanisms are referred to as *autoregulation* (changes in the intracranial vasculature in response to changes in systemic blood pressure) and *chemoregulation* (changes in the intracranial vasculature in response to changes in concentrations of chemical substances in the arterial tree or perivascular spaces).

Acetazolamide (ACTZ) is a carbonic anhydrase inhibitor that causes vasodilation of the intracranial vasculature. Although the mechanism of action is still not completely understood, ACTZ seems to decrease the removal of $CO_2$ from blood and tissue [12] and may also have a direct effect on the vasculature [13, 14]. Acetazolamide does not require patient cooperation and is widely used. However, the effect on the intracranial vasculature is less predictable and reproducible than with $CO_2$, and the injection may have side effects, including arterial hypertension and headaches [15].

Carbon dioxide and oxygen are common and very powerful stimuli influencing the intracranial blood vessels and affecting CBF. The effect of increasing or decreasing concentrations of these substances is vasodilatation or vasoconstriction, affecting intracranial vessel diameter and, consequently, CBF. However, $O_2$ concentrations, contrary to $CO_2$ concentrations, do not have a direct relationship to changes in CBF. A partial pressure of oxygen ($pO_2$) under 50 mmHg is required for changes in CBF to be noticed. This is likely due to the fact that a significant reduction in hemoglobin saturation (arterial oxygen content) does not occur until arterial $pO_2$ falls to levels around 50–60 mmHg, indicating that the actual influence to CBF changes is the arterial content of $O_2$ and not $pO_2$ [16].

Changes in carbon dioxide concentration in the arterial blood ($pCO_2$) have a direct relationship with changes in CBF. It's been well established for decades that increasing levels of $CO_2$ produce vasodilation and decreasing $CO_2$ concentrations lead to vasoconstriction of the distal intracranial circulation [17]. Fast changes in arterial $pCO_2$ lead to proportional changes in CBF by reactive changes in microcirculation vessel diameter and, consequently, cerebrovascular resistance. The diameter of the large vessels at the circle of Willis remains constant during changes in $pCO_2$ [18]. The caliber of the distal arterioles increases with higher levels of $pCO_2$ and decreases with decreasing $pCO_2$, leading to changes in cerebrovascular resistance and, consequently, CBF and CBF velocity, which can be detected by TCD on the large vessels [19]. However, compensatory mechanisms are available to re-establish normal CBF if changes in $CO_2$ are prolonged (hours) [20]. Carbon dioxide seems to have a direct and an indirect, pH-mediated, effect on the distal circulation diameter [21, 22].

It is important to remember that transcranial Doppler does not evaluate CBF directly. It measures CBF velocity (CBFv), and changes in CBF can be inferred from changes in CBFv. Markwalder et al. [15] demonstrated that changes in CBFv accurately reflect changes in CBF if blood pressure remains constant and that the angle between the ultrasound probe and the vessel remains constant. Conditions that influence CBF such as hypertension, vasospasm, anemia, hemodilution, and operator-related factors must be taken into consideration when performing and interpreting TCD measurements of CBFv.

## 4   Carbon Dioxide Manipulation

Many techniques have been described to manipulate $CO_2$ during CVR testing. The choice of $CO_2$ manipulation method will depend on local expertise, clinical condition (e.g., ICU × outpatient), resources and type of information required, and the environment where CVR is being tested. Breath-holding is the simplest way to increase arterial $CO_2$. End-tidal $CO_2$ can be measured, but the degree of change in $PaCO_2$ is not possible to control and may be different between individuals and in the same individual in different tests. Patients are asked to breath normally at room air for a few minutes and then hold their breath after a normal inspiration-expiration cycle. Maximal inspiration induces a Valsalva effect that can affect CBFv measurements by increasing venous resistance. Continuous TCD monitoring of MCA CBFv is used to determine the baseline and its change with increased $CO_2$. To calculate the *breath-holding index*, the percentage increase in MCA CBFv is divided by the breath-holding time in s.

Ringelstein et al. [23] and Bishop et al. [24] described two different techniques for additional $CO_2$ inhalation and CVR testing. When changes in CBFv and $CO_2$ are plotted, a sigmoid curve is obtained reflecting the whole range of cerebrovascular vasomotor reactivity. Ringelstein's method provides an accurate method for CVR evaluation but is time-consuming. The levels of $CO_2$ required can be high and can cause a sensation of asphyxiation, not well tolerated by many individuals.

Bishop's method is simpler, less time-consuming, and better tolerated. Subjects inhale a fixed amount of additional $CO_2$, and the percentage increase in CBFv is divided by the increase in end-tidal $CO_2$ during the test. Although it does not allow for evaluation of the full range of intracranial vascular reactivity, it is better tolerated. The percentage change in CBFv using 5% $CO_2$ inhalation as proposed by Bishop correlated well with changes in CBF using $^{133}$Xenon [24].

The accurate measurement and control of blood levels of $CO_2$ is the main issue with reproducibility of the abovementioned techniques. If changes in $CO_2$ levels are not reported or accurately measured [25, 26], reliability and reproducibility of the test are compromised. To overcome individual variations, methods to control arterial gases based on end-tidal $CO_2$ and $O_2$ measurements and variable concentrations of inhaled $CO_2$ were developed. Slessarev et al. [27] demonstrated that end-tidal pressures of $CO_2$ and $O_2$ can be independently controlled in non-intubated, spontaneously breathing patients using a circuit to prospectively target gas levels. This method has been developed further and used to evaluate cerebrovascular reactivity in many neurological/neurosurgical pathologies, often using MRI (BOLD) techniques [28–34]. Briefly, a rebreathing circuit connected to a computerized gas mixer enables tight control and quick changes of end-tidal $CO_2$ while keeping end-tidal $O_2$ levels constant, independent of the subject's minute ventilation. Reproducible changes in $CO_2$ levels can be induced within three breaths and are accurate to within 1 mmHg [27]. This technique seems to provide the most reliable way of controlling arterial levels of $CO_2$ [35]. The method has also been used in conjunction with TCD to study CVR in patient with traumatic brain injury (TBI) and aSAH [30, 31].

## 5   Cerebrovascular Reactivity

Cerebrovascular reactivity (CVR) can be defined as the change in cerebral blood flow in response to a vasoactive stimulus. The most commonly used stimuli are acetazolamide [36] and $CO_2$ [37]. After the development of methods to measure CBF directly, interest in monitoring cerebrovascular reactivity using various types of stimuli followed. In principle, any method that can measure CBF directly or indirectly can be used to test CVR. The most commonly used methods to measure CBF are $^{133}$Xenon clearance [16], stable xenon-CT (Xe-CT) blood flow measurements [38], positron emission tomography (PET) scan [39], transcranial Doppler [24], magnetic resonance imaging (MRI) methods [40, 41], and near-infrared spectroscopy [42]. Regardless of the method used to measure CBF, estimation of cerebrovascular reactivity relies on detecting changes from baseline on CBF or CBF velocity as a surrogate of CBF after the introduction of a stimulus. Therefore, measurements of pre- and post-stimulation are required.

Ideal methods are capable of relatively quick data acquisition and good anatomical resolution, and the ideal stimuli are controllable, reliable, short-lived, and fast-acting. All methods mentioned above may have advantages and disadvantages in testing cerebrovascular reactivity: the anatomical resolution of MRI adds to the complexity and costs of obtaining an MRI scan, while TCD, although extremely portable, fast, user-friendly, and widely available, does not provide good anatomical information and cannot be performed in up to 25% of patients due to lack of bony windows. Despite that, the portability and real-time information provided by TCD make it a very convenient method to evaluate cerebral blood flow velocity and CVR in many situations, from athletes at the sidelines after a concussion, soldiers in a battlefield hospital, to severely ill ICU patients.

## 6    Transcranial Doppler Measurement of CVR in Traumatic Brain Injury

Traumatic brain injury (TBI) is a major public health concern, a significant cause of morbidity and mortality worldwide, and a source of very significant personal and social burdens [43, 44]. Although the pathophysiology and management of severe traumatic brain injury have been the focus of much of the TBI research, more recently, awareness regarding the consequences of milder injuries (mild TBI—mTBI) has increased significantly. "Mild" traumatic brain injury (mTBI) or concussions comprise up to 75–90% of head injuries, and up to 30% of patients are left with persistent neurocognitive and psychological symptoms [45, 46].

Prevention of secondary injury is a cornerstone of TBI management. Ischemia due to decreased perfusion is a frequent cause of severe secondary injury after TBI. Ischemia lasting more than 4–5 min results in irreversible brain damage, and it is suggested that secondary ischemia can increase neurological damage even in milder TBI [13]. Maintenance of hemodynamic health and CBF is of extreme importance in the management of patients with TBI. Cerebrovascular reactivity to $CO_2$ is one of the two fundamental physiological mechanisms to control CBF, along with cerebral autoregulation. Changes in CBF after head injury and the relationship of impaired CBF control mechanisms, either autoregulation or cerebrovascular reactivity, with TBI outcomes are well known [47–49].

For practical purposes in CVR evaluation, the TBI population can be divided into two groups: mild TBI and moderate/severe TBI. This distinction is mainly done based on the ability of the patients to collaborate with the test and the need or not for mechanical ventilation. While the TCD measurement method is the same, the techniques to manipulate $CO_2$ have to be adapted to patient condition.

Len et al. [50] used breath-holding and TCD to compare CVR between athletes with and without history of recent concussion. Interestingly, they showed that although MCA blood flow velocity is similar at baseline, cerebrovascular reactivity is impaired on subjects with history of a recent concussion. Observations by Lang [51] highlight the importance of intracranial autoregulation and CVR in the management of patients with TBI. When the mechanisms of control of CBF are impaired, blood flow velocity in the middle cerebral artery (a surrogate measure for CBF) was highly dependent on cerebral perfusion pressure (CPP). Therefore, proper evaluation of the physiological mechanism of control of CBF is of utmost importance in patients with head injury and should be used to guide measures to maintain systemic hemodynamics and control intracranial pressure accordingly. This is especially important in patients with severe TBI. The classic methods for evaluation of CBF and hemodynamics are either invasive or cumbersome for daily measurements, especially in ICU patients [52–54]. In this particular condition, TCD's portability and noninvasiveness make it very well-suited to provide continuous, real-time information [55–58].

Ter Minassian et al. [59] demonstrated a good correlation between changes in MCA blood flow velocity and CBF assessed using the difference in arteriovenous content in oxygen in a series of severe TBI patients, showing that the presence of heterogeneous post-traumatic intracranial mass lesions and intracranial hypertension does not affect the ability of TCD to determine CBF changes during CVR testing with $CO_2$ manipulation.

Y. Zurynski et al. [57] found alterations on CBF after head injury in 35 out of 50 patients with severe TBI (Glasgow Coma Scale $\leq 8$). Using daily TCD measurements for CBF, the authors detected vasospasm in 20 and hyperemia in 15 patients. The authors suggest that early changes in CBF velocity may be an indicator of later development of vasospasm or hyperemia. Although these events seemed to correlate with a worse outcome, patients with hyperemia had the worse outcome [58], and ongoing monitoring allowed the distinction between hyperemia and vasospasm. Since the management of these two complications after severe TBI requires very different interventions, proper identification of the underlying issue is very important [60].

## 7   Transcranial Doppler Measurement of CVR in Aneurysmal Subarachnoid Hemorrhage

TCD has been used to monitor the blood flow velocity in patients with aneurysmal SAH since its introduction by Aaslid [3] in the early 1980s. It has become a routine monitoring technique in the management of patients with ruptured aneurysms, aiming at early detection of vasospasm and the timely institution of rescue measures, either medical or interventional [61]. Despite less than ideal sensitivity and specificity for vasospasm diagnosis, the test can be used to triage patients in need for further investigation or closer monitoring [62], with established criteria [63] for interpretation of blood flow velocity changes after aSAH.

Aneurysmal subarachnoid hemorrhage is known to affect cerebral autoregulation (pressure related) and cerebrovascular reactivity (chemical-related vessel diameter changes) even before the onset of vasospasm [64]. Since du Boulay et al. [37] showed evidence of increased vasomotor tone and vascular sensitivity to changes in $CO_2$ in baboons after induced SAH, there has been extensive research on the impact of SAH on the mechanisms of CBF control [65–69]. It has also been suggested that the degree of disturbance correlates with the patient's clinical condition, that recovery of autoregulatory capacity and cerebrovascular reactivity occurred more often in good-grade patients, and that there might be a correlation between impaired CVR and later development of vasospasm and outcomes [68, 70, 71].

Along these lines, the relationship of changes in CVR after aSAH, the development of vasospasm and delayed ischemic neurological deficit (DIND), and aSAH outcomes has been investigated. Studies investigating the use of CVR in patients with aSAH aiming at predicting the development of vasospasm and DIND have produced inconsistent results. W. Hassler et al. [72] found no abnormal CVR to $CO_2$ in a series of 33 patients with aSAH even during vasospasm.

In more recent studies, J. A. Frontera et al. [73] performed serial TCD and CVR to $CO_2$ studies in 20 patients during the first 8 days after aSAH and suggested that impairment of cerebrovascular reactivity to $CO_2$ was strongly correlated with the development of symptomatic vasospasm. E. Carrera et al. [66] performed 83 CVR studies in 34 patients admitted with aSAH in three distinct periods of admission (Days 0–3, 4–7, and 8–10). They found that progressive decline of CVR during admission is a strong predictor of the development of vasospasm and DIND and, furthermore, that the presence of abnormal CVR at any point was associated with DIND with high sensitivity (91%).

We investigated CVR to $CO_2$ in a small series of good-grade aSAH patients [74, 75]. All patients were in good clinical grade at the time of testing, and exams were performed within 7 days of admission using prospective end-tidal targeting previously described by Slessarev et al. [27] We found significant differences in CVR to $CO_2$ between aSAH patients and healthy controls. Patients with aSAH showed impaired CVR both with hypercapnia and the following decrease in $CO_2$ back to normocapnia (Fig. 1). No hypocapnia was induced due to concerns of stimulating vasoconstriction. Despite the significant difference in CVR in the aSAH group, we could not identify a correlation between impaired CVR to $CO_2$ and the development of vasospasm. In this specific group of good clinical grade patients, only one event of DIND occurred.

## 8  Conclusion

The studies addressing the use of TCD to evaluate and monitor CVR to $CO_2$ suggest, despite heterogeneous methods (breath-holding, rebreathing, prospective targeting) and different populations, that this test can be a useful tool to identify patients with little or no cerebrovascular reserve. This can be used to monitor patients closely for vasospasm after aSAH, to manage severe TBI guiding decisions regarding blood pressure levels and appropriate

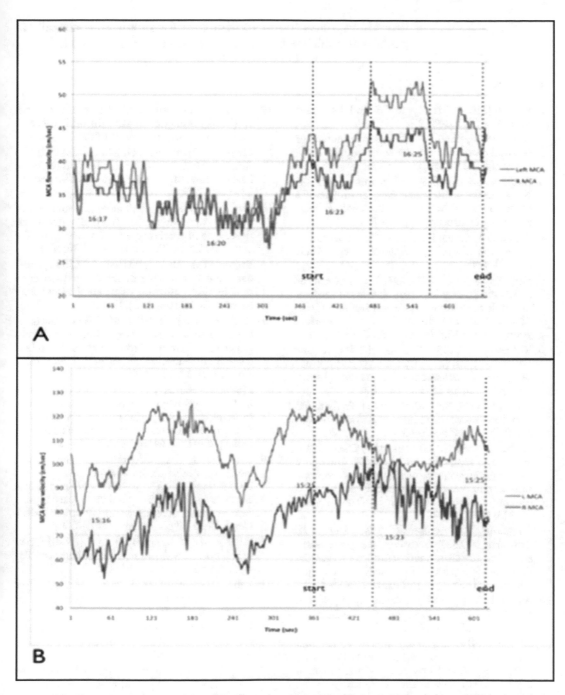

**Fig. 1** (**a, b**) Example of impaired CVR reactivity during $CO_2$ challenge (**b**—aSAH patient). Note the absence of vasomotor response on the RMCA (**b**, purple line) and paradoxical response on the LMCA (**b**, dark blue line)

intracranial pressure management, or to guide return to play after a sports-related concussion. Despite the lack of spatial resolution, the method seems to be able to detect impaired CVR accurately, and its portability and simplicity, not requiring contrast or invasive procedures, make it an excellent tool for monitoring critically ill patients and sideline testing in sports events or in the field.

## References

1. Eden A (1990) The doppler family. Ultrasound Med Biol 16:831–832

2. Miyazaki M, Kato K (1965) Measurement of cerebral blood flow by ultrasonic doppler technique. Jpn Circ J 29:375–382

3. Aaslid R, Markwalder TM, Nornes H (1982) Noninvasive transcranial doppler ultrasound recording of flow velocity in basal cerebral arteries. J Neurosurg 57:769–774

4. Bronk DW, Gesell R (1927) The regulation of respiration. Am J Physiol Leg Cont 82:170–180

5. Kety SS, Schmidt CF (1946) The effects of active and passive hyperventilation on cerebral blood flow, cerebral oxygen consumption, cardiac output, and blood pressure of normal young men. J Clin Invest 25:107–119

6. Kirkpatrick P (1997) Transcranial doppler. In: Peter Reilly RB (ed) Head injury: pathophysiology and management of severe closed head injury. Chapman & Hall, London

7. Purkayastha S, Sorond F (2012) Transcranial doppler ultrasound: technique and application. Semin Neurol 32:411–420

8. Krejza J, Swiat M, Pawlak MA, Oszkinis G, Weigele J, Hurst RW et al (2007) Suitability of temporal bone acoustic window: conventional tcd versus transcranial color-coded duplex sonography. J Neuroimaging 17:311–314

9. Hennerici M, Rautenberg W, Sitzer G, Schwartz A (1987) Transcranial doppler ultrasound for the assessment of intracranial arterial flow velocity--part 1. Examination technique and normal values. Surg Neurol 27:439–448

10. Aaslid R, Lindegaard KF, Sorteberg W, Nornes H (1989) Cerebral autoregulation dynamics in humans. Stroke 20:45–52

11. Khurana VF et al (2004) Biology of cerebral blood vessels and blood flow. In: le Roux PW, Winn HR, Newell DW (eds) Management of cerebral aneurysms. Saunders, Philadelphia, PA

12. Kiss B, Dallinger S, Findl O, Rainer G, Eichler HG, Schmetterer L (1999) Acetazolamide-induced cerebral and ocular vasodilation in humans is independent of nitric oxide. Am J Phys 276:R1661–R1667

13. Ehrenreich DL, Burns RA, Alman RW, Fazekas JF (1961) Influence of acetazolamide on cerebral blood flow. Arch Neurol 5:227–232

14. Hauge A, Nicolaysen G, Thoresen M (1983) Acute effects of acetazolamide on cerebral blood flow in man. Acta Physiol Scand 117:233–239

15. Wolf ME (2015) Functional tcd: regulation of cerebral hemodynamics--cerebral autoregulation, vasomotor reactivity, and neurovascular coupling. Front Neurol Neurosci 36:40–56

16. Brown MM, Wade JP, Marshall J (1985) Fundamental importance of arterial oxygen content in the regulation of cerebral blood flow in man. Brain 108(Pt 1):81–93

17. Kety SS, Schmidt CF (1948) The effects of altered arterial tensions of carbon dioxide and oxygen on cerebral blood flow and cerebral oxygen consumption of normal young men. J Clin Invest 27:484–492

18. Huber P, Handa J (1967) Effect of contrast material, hypercapnia, hyperventilation, hypertonic glucose and papaverine on the diameter of the cerebral arteries. Angiographic determination in man. Investig Radiol 2:17–32

19. Markwalder TM, Grolimund P, Seiler RW, Roth F, Aaslid R (1984) Dependency of blood flow velocity in the middle cerebral artery on end-tidal carbon dioxide partial pressure--a transcranial ultrasound doppler study. J Cereb Blood Flow Metab 4:368–372

20. Raichle ME, Posner JB, Plum F (1970) Cerebral blood flow during and after hyperventilation. Arch Neurol 23:394–403

21. Diji A, Greenfield AD (1960) The local effect of carbon dioxide on human blood vessels. Am Heart J 60:907–914

22. Lambertsen CJ (1960) Carbon dioxide and respiration in acid-base homeostasis. Anesthesiology 21:642–651

23. Ringelstein EB, Sievers C, Ecker S, Schneider PA, Otis SM (1988) Noninvasive assessment of co2-induced cerebral vasomotor response in normal individuals and patients with internal carotid artery occlusions. Stroke 19:963–969

24. Bishop CC, Powell S, Rutt D, Browse NL (1986) Transcranial doppler measurement of

middle cerebral artery blood flow velocity: a validation study. Stroke 17:913–915

25. Dernbach PD, Little JR, Jones SC, Ebrahim ZY (1988) Altered cerebral autoregulation and co2 reactivity after aneurysmal subarachnoid hemorrhage. Neurosurgery 22:822–826

26. Cold GE, Jensen FT, Malmros R (1977) The cerebrovascular co2 reactivity during the acute phase of brain injury. Acta Anaesthesiol Scand 21:222–231

27. Slessarev M, Han J, Mardimae A, Prisman E, Preiss D, Volgyesi G et al (2007) Prospective targeting and control of end-tidal co2 and o2 concentrations. J Physiol 581:1207–1219

28. Conklin J, Fierstra J, Crawley AP, Han JS, Poublanc J, Mandell DM et al (2010) Impaired cerebrovascular reactivity with steal phenomenon is associated with increased diffusion in white matter of patients with moyamoya disease. Stroke 41:1610–1616

29. Conklin J, Fierstra J, Crawley AP, Han JS, Poublanc J, Silver FL et al (2011) Mapping white matter diffusion and cerebrovascular reactivity in carotid occlusive disease. Neurology 77:431–438

30. da Costa L, Fierstra J, Fisher JA, Mikulis DJ, Han JS, Tymianski M (2014) Bold mri and early impairment of cerebrovascular reserve after aneurysmal subarachnoid hemorrhage. J Magn Reson Imaging 40:972

31. da Costa L, van Niftrik CB, Crane D, Fierstra J, Bethune A (2016) Temporal profile of cerebrovascular reactivity impairment, gray matter volumes, and persistent symptoms after mild traumatic head injury. Front Neurol 7:70

32. Fierstra J, Conklin J, Krings T, Slessarev M, Han JS, Fisher JA et al (2011) Impaired perinidal cerebrovascular reserve in seizure patients with brain arteriovenous malformations. Brain 134:100–109

33. Fierstra J, Spieth S, Tran L, Conklin J, Tymianski M, ter Brugge KG et al (2011) Severely impaired cerebrovascular reserve in patients with cerebral proliferative angiopathy. J Neurosurg Pediatr 8:310–315

34. Mandell DM, Han JS, Poublanc J, Crawley AP, Fierstra J, Tymianski M et al (2011) Quantitative measurement of cerebrovascular reactivity by blood oxygen level-dependent mr imaging in patients with intracranial stenosis: preoperative cerebrovascular reactivity predicts the effect of extracranial-intracranial bypass surgery. AJNR. Am J Neuroradiol 32:721–727

35. Fierstra J, Sobczyk O, Battisti-Charbonney A, Mandell DM, Poublanc J, Crawley AP et al (2013) Measuring cerebrovascular reactivity: what stimulus to use? J Physiol 591:5809

36. Yoshida K, Nakamura S, Watanabe H, Kinoshita K (1996) Early cerebral blood flow and vascular reactivity to acetazolamide in predicting the outcome after ruptured cerebral aneurysm. Acta Neurol Scand Suppl 166:131–134

37. Du Boulay G, Symon L, Ackerman RH, Dorsch D, Kendall BE, Shah SH (1973) The reactivity of the spastic arteries. Neuroradiology 5:37–39

38. Gur D, Wolfson SK Jr, Yonas H, Good WF, Shabason L, Latchaw RE et al (1982) Progress in cerebrovascular disease: local cerebral blood flow by xenon enhanced ct. Stroke 13:750–758

39. Gibbs JM, Wise RJ, Leenders KL, Herold S, Frackowiak RS, Jones T (1985) Cerebral haemodynamics in occlusive carotid-artery disease. Lancet 1:933–934

40. Tancredi FB, Gauthier CJ, Madjar C, Bolar DS, Fisher JA, Wang DJ et al (2012) Comparison of pulsed and pseudocontinuous arterial spin-labeling for measuring co2 -induced cerebrovascular reactivity. J Magn Reson Imaging 36:312–321

41. Mark CI, Slessarev M, Ito S, Han J, Fisher JA, Pike GB (2010) Precise control of end-tidal carbon dioxide and oxygen improves bold and asl cerebrovascular reactivity measures. Magn Reson Med 64:749–756

42. Kirkpatrick PJ, Smielewski P, Czosnyka M, Menon DK, Pickard JD (1995) Near-infrared spectroscopy use in patients with head injury. J Neurosurg 83:963–970

43. Frost RB, Farrer TJ, Primosch M, Hedges DW (2013) Prevalence of traumatic brain injury in the general adult population: a meta-analysis. Neuroepidemiology 40:154–159

44. NIH Consensus Development Panel on Rehabilitation of Persons with Traumatic Brain I (1999) Rehabilitation of persons with traumatic brain injury. JAMA 282:974–983

45. Lewine JD, Davis JT, Bigler ED, Thoma R, Hill D, Funke M et al (2007) Objective documentation of traumatic brain injury subsequent to mild head trauma: multimodal brain imaging with meg, spect, and mri. J Head Trauma Rehabil 22:141–155

46. Len TK, Neary JP (2011) Cerebrovascular pathophysiology following mild traumatic brain injury. Clin Physiol Funct Imaging 31:85–93

47. Overgaard J, Tweed WA (1974) Cerebral circulation after head injury. 1. Cerebral blood flow and its regulation after closed head injury with emphasis on clinical correlations. J Neurosurg 41:531–541

48. DeWitt DS, Prough DS (2003) Traumatic cerebral vascular injury: the effects of concussive brain injury on the cerebral vasculature. J Neurotrauma 20:795–825

49. Junger EC, Newell DW, Grant GA, Avellino AM, Ghatan S, Douville CM et al (1997) Cerebral autoregulation following minor head injury. J Neurosurg 86:425–432

50. Len TK, Neary JP, Asmundson GJG, Goodman DG, Bjornson B, Bhambhani YN (2011) Cerebrovascular reactivity impairment after sport-induced concussion. Med Sci Sports Exerc 43:2241–2248

51. Lang EW, Lagopoulos J, Griffith J, Yip K, Yam A, Mudaliar Y et al (2003) Cerebral vasomotor reactivity testing in head injury: the link between pressure and flow. J Neurol Neurosurg Psychiatry 74:1053–1059

52. Kety SS, Schmidt CF (1948) The nitrous oxide method for the quantitative determination of cerebral blood flow in man: theory, procedure and normal values. J Clin Invest 27:476–483

53. Yonas H (1994) Use of xenon and ultrafast ct to measure cerebral blood flow. AJNR. Am J Neuroradiol 15:794–795

54. Joseph MN, Stable JL (2000) Xenon computed tomography cerebral blood flow measurement in neurological disease: review and protocols. Int J Emerg Intens Care Med 4

55. Chan KH, Dearden NM, Miller JD (1993) Transcranial doppler-sonography in severe head injury. Acta Neurochir Suppl (Wien) 59:81–85

56. Ng SCP, Poon WS, Chan MTV, Lam JMK, Lam W, Metreweli C (2000) Transcranial doppler ultrasonography (TCD) in ventilated head injured patients: correlation with stable xenon-enhanced CT. Springer, Vienna, pp 479–482

57. Zurynski YA, Dorsch NW, Pearson I (1995) Incidence and effects of increased cerebral blood flow velocity after severe head injury: a transcranial doppler ultrasound study I. Prediction of post-traumatic vasospasm and hyperemia. J Neurol Sci 134:33–40

58. Zurynski YA, Dorsch NWC, Fearnside MR (1995) Incidence and effects of increased cerebral blood flow velocity after severe head injury: a transcranial doppler ultrasound study II. Effect of vasospasm and hyperemia on outcome. J Neurol Sci 134:41–46

59. Ter Minassian A, Melon E, Leguerinel C, Lodi CA, Bonnet F, Beydon L (1998) Changes in cerebral blood flow during paco2 variations in patients with severe closed head injury: comparison between the fick and transcranial doppler methods. J Neurosurg 88:996–1001

60. Gomez CR, Backer RJ, Bucholz RD (1991) Transcranial doppler ultrasound following closed head injury: vasospasm or vasoparalysis? Surg Neurol 35:30–35

61. Rigamonti A, Ackery A, Baker AJ (2008) Transcranial doppler monitoring in subarachnoid hemorrhage: a critical tool in critical care. Can J Anaesth 55:112–123

62. Creissard P, Proust F, Langlois O (1995) Vasospasm diagnosis: theoretical and real transcranial doppler sensitivity. Acta Neurochir 136:181–185

63. Tsivgoulis GN, Neumyer MM, Alexandrov AV (2011) Diagnostic criteria for cerebrovascular ultrasound. In: Alexandrov A (ed) Cerebrovascular ultrasound in stroke prevention and treatment. Wiley-Blackwell, Singapore

64. Giller CA (1989) Transcranial doppler monitoring of cerebral blood velocity during craniotomy. Neurosurgery 25:769–776

65. Lam JM, Smielewski P, Czosnyka M, Pickard JD, Kirkpatrick PJ (2000) Predicting delayed ischemic deficits after aneurysmal subarachnoid hemorrhage using a transient hyperemic response test of cerebral autoregulation. Neurosurgery 47:819–825. discussions 825–816

66. Carrera E, Kurtz P, Badjatia N, Fernandez L, Claassen J, Lee K et al (2010) Cerebrovascular carbon dioxide reactivity and delayed cerebral ischemia after subarachnoid hemorrhage. Arch Neurol 67:434–439

67. Hashi K, Meyer JS, Shinmaru S, Welch KM, Teraura T (1972) Cerebral hemodynamic and metabolic changes after experimental subarachnoid hemorrhage. J Neurol Sci 17:1–14

68. Ishii R (1979) Regional cerebral blood flow in patients with ruptured intracranial aneurysms. J Neurosurg 50:587–594

69. Soehle M, Czosnyka M, Pickard JD, Kirkpatrick PJ (2004) Continuous assessment of cerebral autoregulation in subarachnoid hemorrhage. Anesth Analg 98:1133–1139. table of contents

70. Abe K, Demizu A, Kamada K, Shimada Y, Sakaki T, Yoshiya I (1992) Prostaglandin e1 and carbon dioxide reactivity during cerebral aneurysm surgery. Can J Anaesth 39:247–252

71. Meixensberger J (1993) Xenon 133--cbf measurements in severe head injury and subarachnoid haemorrhage. Acta Neurochir Suppl (Wien) 59:28–33

72. Hassler W, Chioffi F (1989) Co2 reactivity of cerebral vasospasm after aneurysmal subarachnoid haemorrhage. Acta Neurochir 98:167–175

73. Frontera JA, Rundek T, Schmidt JM, Claassen J, Parra A, Wartenberg KE et al (2006) Cerebrovascular reactivity and vasospasm after subarachnoid hemorrhage: a pilot study. Neurology 66:727–729

74. Da Costa L, Houlden D, Rubenfeld G, Tymianski M, Fisher J, Fierstra J (2015) Impaired cerebrovascular reactivity in the early phase of subarachnoid hemorrhage in good clinical grade patients does not predict vasospasm. Springer International Publishing, Cham, pp 249–253

75. Costa LB (2014) Development of an improved bedside methodology for measurement of cerebrovascular reactivity. Master thesis. University of toronto. Canada

# Chapter 5

# The Role of Cerebrovascular Reactivity Mapping in Functional MRI: Calibrated fMRI and Resting-State fMRI

## J. Jean Chen and Claudine J. Gauthier

## Abstract

Functional MRI (fMRI) is primarily based on the same blood oxygenation level-dependent (BOLD) phenomenon that MRI-based cerebrovascular reactivity (CVR) mapping has relied upon. This technique is finding an ever-increasing role in neuroscience and clinical research as well as treatment planning. The estimation of CVR has unique applications in and associations with fMRI. In particular, CVR estimation is part of a family of techniques called calibrated BOLD fMRI, the purpose of which is to allow the mapping of cerebral oxidative metabolism (CMRO2) using a combination of BOLD and cerebral blood flow (CBF) measurements. Moreover, CVR has recently been shown to be a major source of vascular bias in computing resting-state functional connectivity, in much the same way that it is used to neutralize the vascular contribution in calibrated fMRI. Furthermore, due to the obvious challenges in estimating CVR using gas challenges, a rapidly growing field of study is the estimation of CVR without any form of challenge, including the use of resting-state fMRI for that purpose. This review addresses all of these aspects in which CVR interacts with the fMRI and provides a view into the future of noninvasive CVR measurement.

Keywords Calibrated BOLD, Neurovascular coupling, Cerebrovascular reactivity, Resting-state fMRI, Functional connectivity

## 1  BOLD Signal Physiology

Functional MRI (fMRI) is predominantly performed using the blood oxygenation level-dependent (BOLD) signal. This signal is based on the paramagnetic properties of deoxyhemoglobin, providing a sensitive but unspecific marker of neuronal activity. This lack of specificity stems from the fact that most deoxyhemoglobin (dHb) locally arises from baseline metabolism, with a more modest contribution from task-evoked neuronal activity. The signal measured during a task is due to the dilution of these two sources of dHb from a feedforward cascade of events leading to vasodilation in arterioles, bringing fully oxygenated, and therefore diamagnetic, blood to the area of activity [1, 2]. Therefore, rather than being a direct marker of neuronal activity, the BOLD signal reflects the

Jean Chen and Jorn Fierstra (eds.), *Cerebrovascular Reactivity: Methodological Advances and Clinical Applications*, Neuromethods, vol. 175, https://doi.org/10.1007/978-1-0716-1763-2_5, © Springer Science+Business Media, LLC, part of Springer Nature 2022

relative interplay between baseline oxidative metabolism, task-evoked metabolism, neurovascular coupling mechanisms, and the extent to which local vessels dilate in response to these neurovascular coupling chemical signals [1, 3]. It is this last aspect that underlies the amplitude of the cerebrovascular reactivity (CVR) response measured by BOLD fMRI.

While BOLD-based fMRI is widely used and has several applications in clinical fields [3–6], its physiologically unspecific nature makes it vulnerable to a variety of biases, especially in clinical populations [5, 7–10]. It has been estimated that in healthy brains, the vascular response component is about twice the amplitude of the metabolic response [11, 12]. The vascular response consists of both a blood flow and a blood volume response, but because blood flow has a supralinear dependence on vessel diameter (modeled using Poiseuille's law), it is the blood flow response that dominates the BOLD response. In less healthy populations, changes in vascular elasticity or neurovascular coupling mechanisms can lead to reduced vasodilation [1, 2]. Because of this supralinear dependence on diameter, even small differences in diameter changes with aging or disease can have a large impact on blood flow as compared to young healthy populations. Therefore, the different physiological subcomponents that make up the BOLD signal may not contribute identically to the measured signal in populations of different ages or presenting with different health conditions. This can lead to systematic biases in many BOLD signal comparisons across groups [13].

## 2  The Role of CVR in Calibrated fMRI

There exists a variety of methods to extract or correct the BOLD signal and make it a more quantitative marker of neuronal activity [3, 14]. Notably, CVR is a subcomponent of a family of techniques called calibrated fMRI, which is the predominant approach to quantify and extract the neuronal and vascular components of the BOLD response [12, 15–18]. In its fuller implementations, calibrated fMRI allows the separation of the BOLD signal into its baseline and task-induced vascular and metabolic components [18–20]. In this chapter, the role of CVR in calibrated fMRI will be discussed.

## 3  BOLD Sensitivity to $CO_2$

The physiological mechanisms of vascular reactivity to $CO_2$ have been introduced in Chapters 1–4 and in this chapter. In the past decades, the $CO_2$-driven BOLD response has been the preeminent method for mapping CVR. $CO_2$ is a potent vasodilator used that has been shown to rely mainly on the nitric oxide (NO) pathway to increase arterial diameter [1, 21–23]. While the exact source of NO (endothelial, neuronal, or astrocytic) is still debated, NO production has been shown to mirror changes in $CO_2$ partial pressure, with, for example, 40% increases in $CO_2$ partial pressure resulting in a 36% increase in NO production through endothelial cells in Fathi et al. [24]. Vessel diameter is highly sensitive to the surrounding $CO_2$ concentration, with increasing $CO_2$ partial pressures leading to linear increases in both vessel diameter and flow [25, 26]. In Komori et al., for example, this increase was shown to be 21.6% for arteriolar diameter and 34.5% for flow velocity for a 50% change in $CO_2$ partial pressure in rabbit arterioles [26]. This sensitivity can be captured using MRI, within our data, a 12.0% change in inhaled $CO_2$ concentration resulting in a 24.9% change in gray matter CBF measured using arterial spin labeling (ASL) and a 1.5% change in the gray matter BOLD signal [27]. Since the BOLD signal has both a static and a temporal signal-to-noise ratio (SNR) that is typically above 100 [27, 28], it is a sensitive measure of $CO_2$-induced vasodilation at the whole-brain level.

# 4   Calibrated fMRI

The physiological mechanisms of the vascular response to neuronal activity that engenders the BOLD signal are shown in Fig. 1. In its most common form, calibrated fMRI uses breathing manipulations to estimate the blood flow and blood volume component of the BOLD response to a task, to separate it from the non-vascular component of the BOLD signal [12, 15–18]. The most common calibration procedure for this type of technique uses hypercapnia or increased $CO_2$ concentration in inhaled air, to cause a putatively purely vascular response [12, 16]. This calibration procedure is based on the underlying assumption that $CO_2$, known to be a potent vasodilator, does not cause any change in oxidative metabolism. This vascular $CO_2$ response is essentially CVR.

The original calibrated fMRI model was presented by Davis et al. in 1998 [16], followed in 1999 by a more complete description of the dHb dilution model that underlies it by Hoge et al. [12] In this model, the BOLD signal measured during hypercapnia is related to the CBF signal measured using ASL during hypercapnia, the calibration $M$ parameter, and two other parameters typically assumed from the literature: alpha, which represents flow-volume coupling, and beta, which represents the field strength-dependent magnetic properties of dHb. The BOLD and CBF components can be measured, while alpha and beta are assumed, and $M$ is the output of this calibration procedure. Conceptually, $M$ represents the maximum possible BOLD signal. Since hypercapnia is assumed to be metabolically neutral, then $M$ corresponds to the BOLD signal one

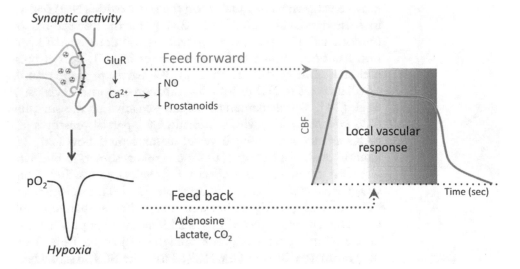

**Fig. 1** Local vascular response evoked by synaptic activity. Glutamate released by synaptic activity activates postsynaptic glutamate receptors (GluRs), leading to activation of calcium-dependent signaling pathways, which results in the release of vasoactive factors at the local capillary/arteriole level, including nitric oxide (NO) and prostanoids. A reduction in tissue $O_2$ (hypoxia) caused by the resultant upregulation of energy consumption leads to the accumulation of metabolic byproducts (adenosine, lactate, and $CO_2$) that also contribute to local vasodilation to better match the perfusion needs of the tissue. (Figure reproduced from [1] with permission from Cell Press)

would obtain if all dHb present in the brain from baseline metabolism were suddenly removed. At 3 T, this value is between 4% and 12% when using a hypercapnia model; see review in [3, 14, 29, 30]. To perform a calibrated fMRI experiment, therefore, one measures the BOLD and CBF percent signal change in response to mild hypercapnia and then uses the calibration equation from the model to extrapolate to the asymptote of the curve, corresponding to this maximal dilution of dHb. CVR is an intermediate measurement of this technique since it is measured as the BOLD or CBF percent change per mmHg change in $CO_2$ concentration during hypercapnia.

The next step of the calibrated fMRI framework per the Davis model is then to estimate the oxidative metabolism component of the BOLD signal measured in response to a task, by combining the $M$ parameter already measured, the BOLD signal measured in response to the task, the CBF signal measured in response to the same task, and the same alpha and beta parameters mentioned in the calibration procedure. These alpha values are assumed to be the same for the task and calibration (essentially CVR) procedures in most cases, though some work has shown that these may be different [31, 32]. It is also noteworthy that some work has suggested that the model should be treated as a heuristic model rather than a biophysical model and that the value of alpha and beta can be determined through data fitting, resulting in a different set of values than what has typically been used in the literature [33].

Other versions of calibrated fMRI have been developed following this initial formulation. These other models are based on other breathing manipulations such as hyperoxia [15] or a combination of hypercapnia and hyperoxia [18]. While hyperoxia-based calibration improves comfort and has been shown to lead to reliable estimates of $M$ and CMRO2, there is evidence that this model underestimates the true $M$ and CMRO2 [18]. Furthermore, this implementation does not allow estimation of CVR as an intermediate byproduct, which may be valuable in several clinical populations. Finally, extensions of these calibrated fMRI models have also been developed to measure metabolism at rest [18–20]. These other techniques have been reviewed elsewhere [3].

## 5  Assumptions

The most common calibrated fMRI technique is based on the important assumption that $CO_2$ inhalation is metabolically neutral. At high doses, $CO_2$ is likely to cause changes in metabolism, but it is typically assumed that the smaller concentrations used in calibrated fMRI (on the order of 5% $CO_2$ in most cases) cause negligible changes in metabolism. This is a debated assumption, however, with some studies showing decreased metabolism during $CO_2$ inhalation [34, 35] and some showing no change [36, 37]. Aside from the difficulty in proving the null hypothesis that $CO_2$ is metabolically neutral, whether a metabolic activity is detected in response to $CO_2$ may be dependent on the technique and the hypercapnia level used to measure it. If hypercapnia does impact oxidative metabolism and thus bias the CVR estimate used in the calibration step, an accurate measurement of this bias is crucial, as it has been shown to have a large impact on the output of the calibrated fMRI model [29]. However, correction of the model to account for the change in CMRO2 would be straightforward should an accurate measurement of this effect arise as a CMRO2 change parameter could be added to the $M$ equation rather than assuming a value of 1 [29, 34].

The Davis model also assumes that arterial blood is fully oxygenated. This is generally a reasonable assumption, as normal oxygen saturation for arterial blood is typically in the range of 97–100% in young subjects [38]. However, this assumption may be problematic in older [39] and diseased populations [40, 41], which could suffer from lower saturations. However, modeling of the effects of anemia has shown that anemia has a very limited impact on the results of the model [29].

Another underlying assumption of this technique is that the chemical signaling that underlies the neurovascular response is comparable to the signaling that underlies $CO_2$-mediated vasodilation. This is because unless these two types of signaling are comparable, then using hypercapnia to assess the vascular response corresponding to a functional task may be misleading and be associated with the very systematic biases between populations that calibrated fMRI was designed to address. Neurovascular coupling is a complex orchestration of signals with many cell types and pathways involved. Detailed studies have shown that when they are active, neurons and interneurons release NO and that this leads to vasodilation at the arteriolar level [1, 42–44]. Capillary dilation is, however, dependent on astrocytic activation of other pathways, especially the arachidonic acid pathway [45]. Nevertheless, it has been shown that inhibition of nNOS leads to an almost complete diminution of the BOLD and CBF response to neuronal stimulation in rats [46], establishing nNOS as one of the main mediators of the neurovascular coupling that underlies the fMRI signals.

The vasodilatory response to hypercapnia on the other hand is predominantly due to activation of the NOS pathway, leading to the release of NO from neurons [21] and endothelial cells [22, 23]. When the NO pathway is blocked, the vasodilatory response is reduced by 36–94% depending on the inhibitor used, hypercapnia levels, and species [47]. While some studies have shown that inhibiting the endothelial NO pathway or destroying endothelial cells does not abolish the CBF response to hypercapnia [48], other studies have shown that neuronal sources cannot in isolation explain the CBF response to hypercapnia [49–51]. This likely reflects a combined contribution of endothelial and neuronal sources or redundancy that allows one system to come online when the other fails. It is important to note, however, that there may be some important species-related differences in pathways [23], making animal results only partly relevant to human data. Therefore, while it is currently unclear whether these two responses are truly equivalent, there are clear similarities between them, lending validity to the use of hypercapnia-based CVR as a model for the vascular component of neurovascular coupling.

## 6    The Role of CVR in Resting-State fMRI

### 6.1    What Is Resting-State fMRI?

fMRI in the resting state (rs-fMRI), particularly based on the BOLD signal, has been extensively used to measure functional connectivity in the brain. The use of the BOLD signal for resting-state imaging largely began with the seminal discovery of resting-state BOLD (rs-BOLD) signal-based synchronization across brain networks (namely, "resting-state functional connectivity") by Biswal et al. [52, 53]. Despite the undefined cognitive state of the brain in the "resting state" and the ambiguous involvement of vascular and metabolic mechanisms underlying the BOLD signal (discussed later in this section), resting-state BOLD signal-based brain networks have been consistently revealed in numerous studies. Notably, the well-documented default-mode network (DMN), the key in generating cognition, has been implicated in a wide array of neurological diseases. In recent years, resting-state BOLD-fMRI has gained considerable attention in basic and clinical neuroscience [52, 53], and the number of publications using the resting-state BOLD contrast has seen exponential growth. This remarkable growth is attributable to the ability of rs-BOLD studies to bypass the hurdles of task performance and behavioral evaluations in assessing brain function, opening a new attractive avenue for neuroimaging research in pediatrics, aging, and a variety of neurologic and psychiatric diseases. A comprehensive description of the applications of rs-BOLD signals can be found in recent reviews [53].

Functional connectivity is the main reason for the popularity of resting-state fMRI. First-level resting-state functional connectivity is generally computed using seed-based correlation or data-driven approaches. In the model-based seed-correlation analysis [52, 54], connectivity is defined as the correlation between the seed rs-fMRI signal time series and those of other brain voxels or regions. Data-driven methods typically use principal or independent component analysis (PCA and ICA, respectively) to identify brain networks. While the seed-based approach is constrained by model assumptions and a priori hypotheses, model-free data-driven methods are more challenging to use in group analyses due to higher variability in the network components identified. Seed- and data-driven approaches yield largely similar spatial patterns (although with differing spatial extents), and both methods can be used to determine connectivity magnitude. However, due to the linear assumptions related to the model-based connectivity methods, it is easier to intuit the influence of physiological metrics on resting-state functional connectivity. After all, there is a well-defined relationship between the correlation coefficient and the rs-fMRI signal amplitude, which is, in turn, describable by the steady-state fMRI signal model as introduced earlier. With the adoption of higher-level functional connectivity measures (those derived from the first-level metrics, such as centrality and hubness, to name a few), the effect of physiological biases may not seem obvious, but it is all the more important to understand them at more abstract levels of analysis.

## 6.2 The Role of CVR in Resting-State fMRI

The rs-fMRI technique, while immensely popular, has been limited by a lack of a fundamental physiological understanding of the underlying rs-fMRI BOLD signal [55]. The BOLD signal is only an indirect measure of neuronal activity and is inherently modulated by both neuronal activity and vascular physiology [56–59]. Currently, the respective contributions of these factors to resting-state BOLD are still unknown. This knowledge gap leads to great challenges for data interpretation in clinical scenarios, whereby these contributions are often altered. The literature suggests that the BOLD-based fMRI signal is fundamentally modulated by local vascular physiology [60–62].

Previous work on the biophysical origins of the rs-fMRI signal suggests that the steady-state BOLD model [12, 16], as outlined in the previous section, is a reasonable framework for understanding the neurovascular underpinnings of the resting BOLD effect. CVR is known to covary with the BOLD response to neuronal activation [63]. Specifically, reduced vascular responsiveness has been associated with reduced BOLD activation amplitude as well as a slowing down in the BOLD response dynamics [64, 65], setting the stage for our study of the effect of CVR on the rs-fMRI signal. Indeed, CVR is a major factor determining the hemodynamic response to neuronal activity, which in turn modulates rs-fMRI signal amplitude (see review [62]). As a result, CVR is expected to drive the amplitude of resting BOLD signal fluctuations (RSFA), as shown in Fig. 2. Indeed, hypercapnia, which elevates basal CBF and oxygenation while reducing CVR [67], has been shown to reduce the amplitude of resting-state BOLD signals [35, 68], consistent with predictions based on the BOLD signal model. Although hypercapnia has also been shown to reduce the power of the alpha rhythm [35], CVR is likely to also play a key role in modulating the RSFA in this context. As an extension, in a study of 335 healthy older adults [69], it was found that the effects of aging on the RSFA were mediated by cardiovascular factors such as heart rate.

However, the relationship between the RSFA and functional connectivity is subtler than linear, as shown in Fig. 3. The hemodynamic response determines the BOLD signal amplitude and subsequently the rs-fMRI BOLD SNR; different signal SNRs will in turn lead to different connectivity measurements [62]. Such biases may obscure the meaning of rs-fMRI functional connectivity measurements [66, 70], which are modulated by the RSFA [65, 70].

In our previous work [70], we demonstrated the extent of this modulation, as well as uncovered the effect of CVR modulation on rs-fMRI functional connectivity. Across the group, rs-fMRI functional connectivity of the motor network also depends significantly on the baseline capnic state, with the hypocapnic baseline associated with the highest connectivity values and hypercapnic baseline associated with the lowest connectivity values. The latter finding is in agreement with early data from Biswal [68] and more recent data from Xu et al. [35]. This association, however, is not consistent across the brain [71]. The recent work by Lewis et al. [72] extends this work to dynamic functional connectivity analyses and over 42 functional networks. It was found that network connectivity is generally weaker during vasodilation, which is supported by previous research [68, 70].

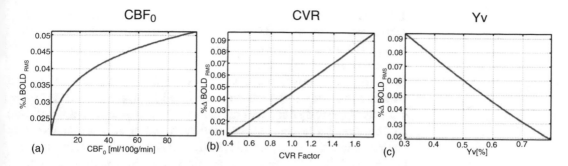

**Fig. 2** The theoretical relationship between rs-fMRI signal amplitude and physiological variables. The BOLD fMRI fluctuation amplitude (%BOLD$_{RMS}$) is plotted against (a) baseline cerebral blood flow (CBF$_0$), (b) cerebrovascular reactivity (CVR), and (c) venous blood oxygenation (Yv). (Figure reproduced from [66] with permission from Elsevier)

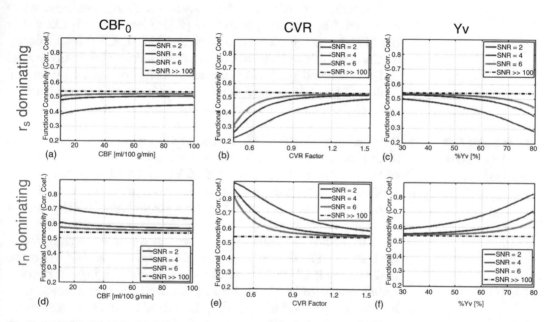

**Fig. 3** (a–f) The theoretical relationship between rs-fMRI functional connectivity and physiological variables. The dependence of functional connectivity (FC) on all three physiological variables (CBF$_0$, CVR, and Yv) is driven by the signal-to-noise ratio (SNR) and by the balance between signal-driven ($r_s$) and noise-driven correlations ($r_n$). A lower SNR leads to a more emphasized dependence of FC on baseline physiology. (Figure reproduced from [66] with permission from Elsevier)

As a follow-up work [66], instead of modulating CVR within individuals, we compared MRI-based functional connectivity (fcMRI) and CVR across different individuals. There was observable inter-subject CVR variation even among healthy young adults, as well as a distribution of functional connectivity values. In this work, we used the steady-state BOLD model to make predictions about functional connectivity given CVR (as well as CBF and SO$_2$). We further used these predictions to help interpret the empirical data, to improve our understanding of the possible origins of inter-subject variations in rs-fMRI functional connectivity. However, the characterization of vascular biases on rs-fMRI metrics remains scarce in current literature.

Given the above, the open question is how to normalize or calibrate the vascular effects from rs-fMRI measures, which is especially used to study neuronal markers [73–75]. As an example of the application of a normalization approach, Xu et al. normalized the resting-state BOLD amplitudes in the default-mode network by the task-related BOLD responses in the visual cortex, as $CO_2$ challenges were observed not to alter the RSFA in the visual cortex [35]. More recently, following the findings in [66], Champagne et al. normalized functional connectivity by local CBF and observed a reduction in connectivity differences between healthy and patient groups [73]. This is in line with observations by Garrett et al., who also observed age-related differences in RSFA to be reduced when normalized by vascular physiology [76]. However, it is likely that the MRI-based functional connectivity (fcMRI)-CVR association is mechanical (driven by local CVR and the local vessel composition) and coincidental (driven by the relationship of both to neuronal health). Hence, adequate correction of vascular bias remains an open challenge.

### 6.3 Mapping CVR Without Breathing Challenges

Resting-state fMRI offers a unique opportunity to glean CVR information without the need for respiratory challenges. This type of "unconstrained" or "task-free" CVR protocol does not require cooperation from participants and is thus a promising direction of research that will likely broaden the accessibility of CVR mapping to clinical researchers [6].

The RSFA was initially introduced as a vascular scaling factor for task-based BOLD responses by Kannurpatti et al. [77], as it was used to scale fMRI task responses in the study of aging. Since then, resting-state methods that do not require $CO_2$ perturbation have flourished [70, 79, 80]. Notably, Kannurpatti et al. [78] reported a comparison of the resting-state fMRI fluctuation amplitude (voxel-wise temporal standard deviation) as a CVR surrogate. Liu et al. introduced a method that uses the low-frequency range of the rs-fMRI signal, regressed against the global signal, to generate a qualitative CVR estimate [79]. The same year, Jahanian et al. introduced the concept of either using the voxel-wise regression coefficients with cerebrospinal fluid signal (measured using rs-fMRI) or using the voxel-wise coefficient of variation to approximate CVR [80]. The former is more akin to the approach of Liu et al., while the latter follows the logic of the method by Kannurpatti et al. In the meantime, the Liu method has also been demonstrated in delineating global and local CVR deficits (in Moyamoya disease and acute stroke, respectively) [81].

While all of these methods have demonstrated correlations with $CO_2$-based CVR measures, they provide qualitative measures of CVR, which complicates inter-subject comparisons, and quantitative CVR mapping techniques using rs-fMRI remain scarce. The only such technique takes advantage of spontaneous fluctuations in end-tidal $CO_2$ while eliminating the effects of heart rate and respiratory volume variability on the fMRI signal [70]. Once the $CO_2$-related BOLD signal is isolated, a deconvolution is performed between the resting-state fMRI and $CO_2$ time courses, and the area under the response function is by definition the quantitative CVR. Thus far, the quantitative CVR method has been found to agree more strongly with the standard gas challenge-based method for characterizing within-subject CVR variations across the brain. Additionally, beyond the magnitude of CVR, the dynamic features of the fMRI response function that are available through this method can also provide useful information. A slowing of the CVR response has been shown to characterize vascular lesions (Poublanc et al. [82]), adding a dimension to the utility of CVR mapping. Indeed, differences between young and older adults have been demonstrated using simply the temporal features of the resting $CO_2$ response function [83].

# 7 Conclusions

CVR is important for the interpretation of both task-based and resting-state fMRI results. The need to incorporate CVR into fMRI data interpretation is increasingly recognized, but more accessible ways of mapping CVR are necessary for wide adoption.

# Acknowledgments

The writing of this review was supported by the Canadian Institutes of Health Research (CIHR), the Heart and Stroke Foundation, and the Michal and Renata Hornstein Chair in Cardiovascular Imaging.

# References

1. Iadecola C (2017) The neurovascular unit coming of age: a journey through neurovascular coupling in health and disease. Neuron 96:17–42

2. Girouard H, Iadecola C (2006) Neurovascular coupling in the normal brain and in hypertension, stroke, and Alzheimer disease. J Appl Physiol 100:328–335

3. Gauthier CJ, Fan AP (2019) BOLD signal physiology: models and applications. NeuroImage 187:116–127

4. Specht K (2019) Current challenges in translational and clinical fMRI and future directions. Front Psychiatry 10:924

5. Chen J, Functional J (2019) MRI of brain physiology in aging and neurodegenerative diseases. NeuroImage 187:209–225

6. Chen JJ (2018) Cerebrovascular-reactivity mapping using MRI: considerations for Alzheimer's disease. Front Aging Neurosci 10:170

7. Ances B, Vaida F, Ellis R, Buxton R (2011) Test-retest stability of calibrated BOLD-fMRI

in HIV- and HIV+ subjects. NeuroImage 54: 2156–2162

8. De Vis JB et al (2015) Age-related changes in brain hemodynamics: A calibrated MRI study. Hum Brain Mapp 36:3973–3987

9. Lajoie I et al (2017) Application of calibrated fMRI in Alzheimer's disease. Neuroimage Clin 15:348–358

10. Mazerolle P et al (2018) Oncological and functional outcomes of trans-oral robotic surgery for pyriform sinus carcinoma: a French GET-TEC group study. Oral Oncol 86:165–170

11. Uludağ K et al (2004) Coupling of cerebral blood flow and oxygen consumption during physiological activation and deactivation measured with fMRI. NeuroImage 23: 148–155

12. Hoge RD et al (1999) Investigation of BOLD signal dependence on cerebral blood flow and oxygen consumption: the deoxyhemoglobin dilution model. Magn Reson Med 42:849–863

13. Gauthier CJ et al (2013) Age dependence of hemodynamic response characteristics in human functional magnetic resonance imaging. Neurobiol Aging 34:1469. https://doi.org/10.1016/j.neurobiolaging.2012.11.002

14. Hoge RD (2012) Calibrated fMRI. Neuroimage 62:930–937

15. Chiarelli PA, Bulte DP, Wise R, Gallichan D, Jezzard P (2007) A calibration method for quantitative BOLD fMRI based on hyperoxia. NeuroImage 37:808–820

16. Davis TL, Kwong KK, Weisskoff RM, Rosen BR (1998) Calibrated functional MRI: mapping the dynamics of oxidative metabolism. Proc Natl Acad Sci U S A 95:1834–1839

17. Hoge RD et al (1999) Linear coupling between cerebral blood flow and oxygen consumption in activated human cortex. Proc Natl Acad Sci U S A 96:9403–9408

18. Gauthier CJ, Hoge RD (2012) Magnetic resonance imaging of resting OEF and CMRO2 using a generalized calibration model for hypercapnia and hyperoxia. NeuroImage 60: 1212–1225

19. Bulte DP et al (2012) Quantitative measurement of cerebral physiology using respiratory-calibrated MRI. NeuroImage 60:582–591

20. Wise RG, Harris AD, Stone AJ, Murphy K (2013) Measurement of OEF and absolute CMRO2: MRI-based methods using interleaved and combined hypercapnia and hyperoxia. NeuroImage 83:135–147

21. Pelligrino DA, Santizo RA, Wang Q (1999) Miconazole represses CO(2)-induced pial arteriolar dilation only under selected circumstances. Am J Phys 277:H1484–H1490

22. Peebles KC et al (2008) Human cerebral arteriovenous vasoactive exchange during alterations in arterial blood gases. J Appl Physiol 105:1060–1068

23. Najarian T et al (2000) Prolonged hypercapnia-evoked cerebral hyperemia via K(+) channel- and prostaglandin E(2)-dependent endothelial nitric oxide synthase induction. Circ Res 87: 1149–1156

24. Fathi AR et al (2011) Carbon dioxide influence on nitric oxide production in endothelial cells and astrocytes: cellular mechanisms. Brain Res 1386:50–57

25. Hülsmann WC, Dubelaar ML (1988) Aspects of fatty acid metabolism in vascular endothelial cells. Biochimie 70:681–686

26. Komori M et al (2007) Permissive range of hypercapnia for improved peripheral microcirculation and cardiac output in rabbits. Crit Care Med 35:2171–2175

27. Gauthier CJ, Hoge RD (2013) A generalized procedure for calibrated MRI incorporating hyperoxia and hypercapnia. Hum Brain Mapp 34:1053–1069

28. Triantafyllou C et al (2005) Comparison of physiological noise at 1.5 T, 3 T and 7 T and optimization of fMRI acquisition parameters. NeuroImage 26:243–250

29. Blockley NP, Griffeth VEM, Stone AJ, Hare HV, Bulte DP (2015) Sources of systematic error in calibrated BOLD based mapping of baseline oxygen extraction fraction. NeuroImage 122:105–113

30. Mark CI, Mazerolle EL, Chen JJ (2015) Metabolic and vascular origins of the BOLD effect: implications for imaging pathology and resting-state brain function. J Magn Reson Imaging 42:231–246

31. Chen JJ, Pike GB (2009) BOLD-specific cerebral blood volume and blood flow changes during neuronal activation in humans. NMR Biomed 22:1054–1062

32. Chen JJ, Pike GB (2010) MRI measurement of the BOLD-specific flow–volume relationship during hypercapnia and hypocapnia in humans. NeuroImage 53:383

33. Griffeth VEM, Buxton RB (2011) A theoretical framework for estimating cerebral oxygen metabolism changes using the calibrated-BOLD method: modeling the effects of blood volume distribution, hematocrit, oxygen extraction fraction, and tissue signal properties on the BOLD signal. NeuroImage 58: 198–212

34. Driver ID, Wise RG, Murphy K (2017) Graded hypercapnia-calibrated BOLD: beyond the iso-metabolic hypercapnic assumption. Front Neurosci 11:276

35. Xu F et al (2011) The influence of carbon dioxide on brain activity and metabolism in conscious humans. J Cereb Blood Flow Metab 31:58–67

36. Chen JJ, Pike GB (2010) Global cerebral oxidative metabolism during hypercapnia and hypocapnia in humans: implications for BOLD fMRI. J Cereb Blood Flow Metab 30: 1094–1099

37. Jain V et al (2011) Rapid magnetic resonance measurement of global cerebral metabolic rate of oxygen consumption in humans during rest and hypercapnia. J Cereb Blood Flow Metab 31:1504–1512

38. Barratt-Boyes BG, Wood EH (1957) The oxygen saturation of blood in the venae cavae, right-heart chambers, and pulmonary vessels of healthy subjects. J Lab Clin Med 50:93–106

39. Hardie JA, Vollmer WM, Buist AS, Ellingsen I, Mørkve O (2004) Reference values for arterial blood gases in the elderly. Chest 125: 2053–2060

40. Cukic V (2014) The changes of arterial blood gases in COPD during four-year period. Mediev Archaeol 68:14–18

41. Slowik JM, Collen JF (2020) Obstructive sleep apnea. StatPearls Publishing, Treasure Island, FL

42. Faraci FM, Brian JE Jr (1994) Nitric oxide and the cerebral circulation. Stroke 25:692–703

43. Attwell D et al (2010) Glial and neuronal control of brain blood flow. Nature 468:232–243

44. Rancillac A et al (2006) Glutamatergic control of microvascular tone by distinct GABA neurons in the cerebellum. J Neurosci 26: 6997–7006

45. Mishra A et al (2016) Astrocytes mediate neurovascular signaling to capillary pericytes but not to arterioles. Nat Neurosci 19:1619–1627

46. Stefanovic B, Schwindt W, Hoehn M, Silva AC (2007) Functional uncoupling of hemodynamic from neuronal response by inhibition of neuronal nitric oxide synthase. J Cereb Blood Flow Metab 27:741–754

47. Iadecola C, Zhang F (1994) Nitric oxide-dependent and -independent components of cerebrovasodilation elicited by hypercapnia. Am J Phys 266:R546–R552

48. Wang Q, Pelligrino DA, Koenig HM, Albrecht RF (1994) The role of endothelium and nitric oxide in rat pial arteriolar dilatory responses to CO2 in vivo. J Cereb Blood Flow Metab 14: 944–951

49. Iadecola C, Zhang F, Xu X (1993) Role of nitric oxide synthase-containing vascular nerves in cerebrovasodilation elicited from cerebellum. Am J Phys 264:R738–R746

50. Iadecola C, Zhang F (1996) Permissive and obligatory roles of NO in cerebrovascular responses to hypercapnia and acetylcholine. Am J Phys 271:R990–R1001

51. Iadecola C, Arneric SP, Baker HD, Tucker LW, Reis DJ (1987) Role of local neurons in cerebrocortical vasodilation elicited from cerebellum. Am J Phys 252:R1082–R1091

52. Biswal B, Zerrin Yetkin F, Haughton VM, Hyde JS (1995) Functional connectivity in the motor cortex of resting human brain using echo-planar mri. Magn Reson Med 34: 537–541

53. Fox MD (2010) Clinical applications of resting state functional connectivity. Front Syst Neurosci 4:19. https://doi.org/10.3389/fnsys.2010.00019

54. Van Dijk KRA et al (2010) Intrinsic functional connectivity as a tool for human connectomics: theory, properties, and optimization. J Neurophysiol 103:297–321

55. Leopold DA, Maier A (2012) Ongoing physiological processes in the cerebral cortex. NeuroImage 62:2190–2200

56. Biswal BB, Kannurpatti SS, Rypma B (2007) Hemodynamic scaling of fMRI-BOLD signal: validation of low-frequency spectral amplitude as a scalability factor. Magn Reson Imaging 25: 1358–1369

57. Biswal BB, Kannurpatti SS (2009) Resting-state functional connectivity in animal models: modulations by exsanguination. Methods Mol Biol 489:255–274

58. Kannurpatti SS, Biswal BB, Kim YR, Rosen BR (2008) Spatio-temporal characteristics of low-frequency BOLD signal fluctuations in isoflurane-anesthetized rat brain. NeuroImage 40:1738–1747

59. Tong Y, Frederick BD (2010) Time lag dependent multimodal processing of concurrent fMRI and near-infrared spectroscopy (NIRS) data suggests a global circulatory origin for low-frequency oscillation signals in human brain. NeuroImage 53:553–564

60. Carusone LM, Srinivasan J, Gitelman DR, Mesulam M-M, Parrish TB (2002) Hemodynamic response changes in cerebrovascular disease: implications for functional MR imaging. AJNR Am J Neuroradiol 23:1222–1228

61. Kannurpatti SS, Motes MA, Rypma B, Biswal BB (2010) Neural and vascular variability and the fMRI-BOLD response in normal aging. Magn Reson Imaging 28:466–476

62. Liu TT (2013) Neurovascular factors in resting-state functional MRI. NeuroImage 80:339–348

63. Stefanovic B, Warnking JM, Rylander KM, Pike GB (2006) The effect of global cerebral vasodilation on focal activation hemodynamics. NeuroImage 30:726–734

64. Behzadi Y, Liu TT (2005) An arteriolar compliance model of the cerebral blood flow response to neural stimulus. NeuroImage 25:1100–1111

65. Rack-Gomer AL, Liu TT (2012) Caffeine increases the temporal variability of resting-state BOLD connectivity in the motor cortex. NeuroImage 59:2994–3002

66. Chu PPW, Golestani AM, Kwinta JB, Khatamian YB, Chen JJ (2018) Characterizing the modulation of resting-state fMRI metrics by baseline physiology. NeuroImage 173:72. https://doi.org/10.1016/j.neuroimage.2018.02.004

67. Cohen ER, Ugurbil K, Kim SG (2002) Effect of basal conditions on the magnitude and dynamics of the blood oxygenation level-dependent fMRI response. J Cereb Blood Flow Metab 22:1042–1053

68. Biswal B, Hudetz AG, Yetkin FZ, Haughton VM, Hyde JS (1997) Hypercapnia reversibly suppresses low-frequency fluctuations in the human motor cortex during rest using echo-planar MRI. J Cereb Blood Flow Metab 17:301–308

69. Tsvetanov KA et al (2015) The effect of ageing on fMRI: correction for the confounding effects of vascular reactivity evaluated by joint fMRI and MEG in 335 adults. Hum Brain Mapp 36:2248–2269

70. Golestani AM, Kwinta JB, Strother SC, Khatamian YB, Chen JJ (2016) The association between cerebrovascular reactivity and resting-state fMRI functional connectivity in healthy adults: the influence of basal carbon dioxide. NeuroImage 132:301–313

71. Tak S, Polimeni JR, Wang DJJ, Yan L, Chen JJ (2015) Associations of resting-state fMRI functional connectivity with flow-BOLD coupling and regional vasculature. Brain Connect 5:137–146

72. Lewis N et al (2020) Static and dynamic functional connectivity analysis of cerebrovascular reactivity: an fMRI study. Brain Behav 10:e01516

73. Champagne AA et al (2020) Multi-modal normalization of resting-state using local physiology reduces changes in functional connectivity patterns observed in mTBI patients. Neuroimage Clin 26:102204

74. Coverdale NS, Fernandez-Ruiz J, Champagne AA, Mark CI, Cook DJ (2020) Co-localized impaired regional cerebrovascular reactivity in chronic concussion is associated with BOLD activation differences during a working memory task. Brain Imaging Behav 14:2438. https://doi.org/10.1007/s11682-019-00194-5

75. Tsvetanov KA et al (2020) The effects of age on resting-state BOLD signal variability is explained by cardiovascular and cerebrovascular factors. Psychophysiology:e13714

76. Garrett DD, Lindenberger U, Hoge RD, Gauthier CJ (2017) Age differences in brain signal variability are robust to multiple vascular controls. Sci Rep 7:10149

77. Kannurpatti SS, Motes MA, Rypma B, Biswal BB (2011) BOLD signal change: minimizing vascular contributions by resting-state-fluctuation-of-amplitude scaling. Hum Brain Mapp 32:1125–1140

78. Kannurpatti SS, Motes MA, Biswal BB, Rypma B (2014) Assessment of unconstrained cerebrovascular reactivity marker for large age-range FMRI studies. PLoS One 9:e88751

79. Liu P et al (2017) Cerebrovascular reactivity mapping without gas challenges. NeuroImage 146:320–326

80. Jahanian H et al (2017) Measuring vascular reactivity with resting-state blood oxygenation level-dependent (BOLD) signal fluctuations: a potential alternative to the breath-holding challenge? J Cereb Blood Flow Metab 37:2526–2538

81. Taneja K et al (2019) Evaluation of cerebrovascular reserve in patients with cerebrovascular diseases using resting-state MRI: a feasibility study. Magn Reson Imaging 59:46–52

82. Poublanc J, Crawley AP, Sobczyk O, Montandon G, Sam K, Mandell DM, Dufort P, Venkatraghavan L, Duffin J, Mikulis DJ, Fisher JA (2015) Measuring cerebrovascular reactivity: the dynamic response to a step hypercapnic stimulus. J Cereb Blood Flow Metab 35:1746–1756

83. Esmaelbeigi A, Chen JJ (2021) The effect of age on resting state fMRI carbon dioxide response function. Proc Org Hum Brain Map 2019:1403

# Chapter 6

# Hemodynamic Evaluation of Paradoxical Blood Oxygenation Level-Dependent Cerebrovascular Reactivity with Transcranial Doppler and MR Perfusion in Patients with Symptomatic Cerebrovascular Steno-occlusive Disease

## Christiaan Hendrik Bas van Niftrik, Martina Sebök, Giovanni Muscas, Aimée Hiller, Matthias Halter, Susanne Wegener, Luca Regli, and Jorn Fierstra

## Abstract

**Background and Purpose:** In patients with steno-occlusive disease, paradoxical blood oxygenation level-dependent fMRI cerebrovascular reactivity (BOLD-CVR) is a feasible surrogate marker for hemodynamic impairment. BOLD-CVR, however, does not measure hemodynamic changes directly; hence, we study complementary hemodynamic features in brain areas exhibiting paradoxical BOLD-CVR using perfusion-weighted MRI (PW-MRI) and transcranial Doppler (TCD).

**Methods:** Twenty participants with unilateral symptomatic chronic cerebrovascular steno-occlusive disease and ipsilateral paradoxical BOLD-CVR were studied. The region with paradoxical BOLD-CVR was used as a region of interest for the PW-MRI-weighted images. As a comparison, a contralateral analysis was done. Ipsilateral and contralateral TCD flow velocities of the posterior circulation were compared as an indicator of collateral supply.

**Results:** Brain tissue exhibiting paradoxical BOLD-CVR showed prolonged mean transit time and time-to-peak with increased cerebral blood volume. CBF followed a post-stroke time evolution. The ipsilateral posterior cerebral artery (PCA)-P2 segment flow velocity was significantly increased compared to the contralateral side, correlating strongly with the paradoxical BOLD-CVR brain tissue volume.

**Conclusions:** In symptomatic steno-occlusive patients, brain areas with ipsilateral paradoxical BOLD-CVR show clear hemodynamic changes, i.e., prolonged mean transit time and time-to-peak and increased cerebral blood volume. Moreover, increased flow velocities in the ipsilateral PCA-P2 segment indicate increased hemodynamic efforts due to increased need for collateral supply from the posterior circulation. This further supports the premise that paradoxical BOLD-CVR within the symptomatic hemisphere is a reliable surrogate for exhausted perfusion reserve and should be studied in an independent study cohort to determine its value for predicting stroke risk.

**Keywords** Perfusion reserve, Hemodynamic failure, BOLD fMRI, Transcranial Doppler, MR perfusion, Cerebrovascular reactivity, Stroke

Jean Chen and Jorn Fierstra (eds.), *Cerebrovascular Reactivity: Methodological Advances and Clinical Applications*, Neuromethods, vol. 175, https://doi.org/10.1007/978-1-0716-1763-2_6, © Springer Science+Business Media, LLC, part of Springer Nature 2022

# 1    Introduction

The key feature of severe hemodynamic impairment in symptomatic patients with cerebrovascular steno-occlusive disease is a paradoxical (i.e., negative) cerebral blood flow (CBF) response to a vasoactive stimulus, termed exhausted perfusion reserve capacity [1–5]. Consequently, these brain regions also suffer from other altered hemodynamic features including prolonged mean transit time (MTT) and time-to-peak (TTP) and an increased cerebral blood volume (CBV), indicating insufficient collateral pathways in order to maintain adequate perfusion [6–9].

Alternatively, since routinely measuring perfusion reserve remains cumbersome, paradoxical blood oxygenation level-dependent fMRI cerebrovascular reactivity (BOLD-CVR) has been suggested as a feasible surrogate imaging marker [2]. The BOLD-fMRI signal, however, encompasses the relative difference in deoxyhemoglobin in response to carbon dioxide ($CO_2$) to measure CVR and not absolute changes in cerebral blood flow (CBF). Thus, BOLD-CVR only partially depends on changes in CBF, and inconsistencies between paradoxical BOLD-CVR and perfusion reserve have been reported [10].

We therefore study complementary features of hemodynamic impairment in brain areas exhibiting paradoxical BOLD-CVR using perfusion-weighted MRI (PW-MRI) and transcranial Doppler (TCD) in patients with unilateral symptomatic cerebrovascular steno-occlusive disease.

# 2    Methods

## 2.1    Patient Cohort

All participants were selected from an ongoing prospective database of patients undergoing BOLD-CVR studies. An ethical approval was obtained from the local institutional ethical review board (KEK-ZH-Nr. 2012-0427), and all participants had signed a written informed consent before participating. Datasets of participants were selected based on the presence of symtpomatic unilateral cerebrovascular steno-occlusive disease who underwent perfusion-weighted MRI (PW-MRI) and BOLD-CVR imaging within 4 weeks (mean $3 \pm 9$ days) and an intracranial duplex investigation (TCD) at our site within 7 days (mean $2 \pm 4$ days) of BOLD-CVR scanning. Symptomatic cerebrovascular steno-occlusive disease was defined as the experience of neurological deficits caused by a radiologically defined stroke or a transient ischemic attack (TIA) ipsilateral to a radiologically and duplex-defined persisting cerebrovascular steno-occlusive disease. Patients with contralateral vascular disease (i.e., occlusions or stenosis of the internal carotid artery >40%, occlusion or stenosis of the middle cerebral artery) or any pathology located in the posterior circulation were excluded. Other exclusion criteria for this study were new neurological symptoms between the two scans and surgical, neuro-interventional, or medical revascularization between any of the imaging modalities. Vascular risk factors and the presence or absence of a fetal posterior communicating artery variant of each participant were reviewed through medical chart review.

**2.2 Duplex Acquisition**

In our institute, TCD measurements of all intracranial arteries (anterior cerebral artery, posterior cerebral artery, middle cerebral artery) are part of the clinical routine and assessed in all stroke patients. Because of the inclusion of a cohort with different vascular pathologies leading to stroke in the anterior cerebral circulation, we specifically included ultrasound examination of only the posterior circulation, while it is a known parameter of leptomeningeal collateralization, and an increasing flow velocity of arteries in the posterior circulation is an indicator of increased stroke risk [8]. The following arterial segments were routinely analyzed: the ipsilateral (i.e., symptomatic hemisphere) and contralateral posterior cerebral artery (PCA) P1 and P2 segments were examined, and the peak systolic velocities (PSV) and end-diastolic velocities (EDV) were recorded.

**2.3 MRI Acquisition and Processing**

*2.3.1 BOLD Cerebrovascular Reactivity Maps*

All MRI data were acquired on a 3 Tesla Skyra VD13 (Siemens Healthcare, Erlangen, Germany) with a 32-channel head coil. BOLD fMRI scans were acquired with the following parameters: axial two-dimensional (2D) single-shot echo planar imaging sequence planned on the anterior commissure-posterior commissure line plus 20° (on a sagittal image), voxel size for the BOLD fMRI scans $3 \times 3 \times 3$ mm$^3$, acquisition matrix $64 \times 64 \times 35$ ascending interleaved slice acquisition, slice gap 0.3 mm, GRAPPA factor 2 with 32 ref. lines, repetition time (TR)/echo time (TE) 2000/30 ms, flip angle 85°, bandwidth 2368 Hz/Px, and a field of view $192 \times 192$ mm$^2$.

For overlay purposes, we acquired subsequently a high-resolution three-dimensional (3D) T1-weighted magnetization-prepared rapid acquisition gradient echo image with the same orientation as the BOLD fMRI scans. The acquisition parameters of the T1 imaging were voxel size $0.8 \times 0.8 \times 1.0$ mm$^3$ with a field of view $230 \times 230 \times 176$ mm and scan matrix $288 \times 288 \times 176$, TR/TE/TI 2200/5.14/900 ms, and flip angle 8°.

The $CO_2$ stimulus was administered with a computer-controlled gas blender with prospective gas-targeting algorithms (RespirAct™, Thornhill Research Institute, Toronto, Canada). This stimulus allows precise targeting of the arterial partial pressure of oxygen and $CO_2$ [11]. We controlled the $CO_2$ at the subject's own resting $CO_2$ value [12]. During the CVR study, $CO_2$ was increased an average $9.3 \pm 1.9$ mmHg above their resting $CO_2$

value for 80 s. Oxygen was maintained at a level of ~105 mmHg throughout the protocol.

BOLD fMRI volumes were processed using SPM 12 (Statistical Parameter Mapping Software, Wellcome Department of Imaging Neuroscience, University College London, London, UK) and in-house scripts written in Matlab R2019a (The MathWorks Inc., Natick, MA, USA). Pre-processing of the BOLD fMRI datasets included time and motion correction and smoothing with a Gaussian kernel of 6 mm full width at half-maximum. CVR was calculated from the slope of a linear least square fit of the BOLD signal time course to the $CO_2$ time series over the whole time series [13]. This method has been described and applied in previous work concerning stroke patients [2, 14, 15].

### 2.3.2 Perfusion-Weighted MRI

The sequence parameters of the dynamic susceptibility contrast (DSC) T2* perfusion included total time of acquisition (TA) 1:46 min/s, 60 volumes, field of view (FoV) $220 \times 220$ mm$^2$, 25 transversal slices of 4.0 mm thickness with a 30% distance factor, phase-encoding direction A $\gg$ P, voxel size $1.7 \times 1.7 \times 4.0$ mm$^3$, single-shot acquisition resolution $128 \times 128$, GRAPPA factor 3, 60 reference lines, TR/TE 1600/27.0 ms, flip angle 90°, fat suppression, standard B0 shim, True-Form B1 shim, echo spacing 0.94 ms, and receive bandwidth 1220 Hz/Px. Estimation of the perfusion maps—relative CBF, relative CBV, mean transit time (MTT), and time-to-peak (TTP)—was performed using the OLEA Sphere software (Version 3.0 SP6, Olea Medical SA, La Ciotat, France). These maps, defined under the standard perfusion model hypothesis, are derived using automatic vascular input determination and deconvolution.

### 2.4 Paradoxical BOLD-CVR Characterization

The primary goal was to assess the extent of paradoxical BOLD-CVR in a multimodal imaging fashion. All BOLD and PW-MRI volumes were processed in Matlab R2019a. To reduce the number of artificial negative BOLD values caused by susceptibility artifacts, voxels around the anterior cranial fossa were removed. The extent of negative BOLD-CVR values was identified by determining the volume of paradoxical BOLD-CVR on the T1-weighted image. To remove imaging acquisition noise, a cutoff value of $-0.02\%$ CVR was chosen (i.e., a correlation coefficient below $-0.125$ [16]). The region of paradoxical BOLD-CVR was used as an ROI on the PW-MRI-derived maps (CBF, CBV, MTT, TTP). As a comparison, the region of paradoxical BOLD-CVR was mirrored contralaterally for a similar analysis. For the contralateral analysis, only voxels within the white and gray matter (probability 0.9) were analyzed.

To evaluate signs of hemodynamic failure, an ipsilateral contra-lateral comparison was performed (i.e., comparison of the quantitative values of PW-MRI within the region of paradoxical BOLD-CVR to those in the same region on the contralateral hemisphere).

For TCD, flow velocity differences between the ipsi- and contralateral hemispheres were determined.

**2.5 Statistical Analysis**

A normal distribution was assessed visually, as well as using the Shapiro-Wilk test. BOLD-CVR, flow velocity, and volumetric measurements were deemed not normally distributed, and because of the inclusion of only 20 subjects, we treated all variables as non-normally distributed. Therefore, we compared the ipsilateral and contralateral hemispheres using the Wilcoxon rank sum test. The two-tailed significance was based on $p < 0.05$. Correlations between the volume of paradoxical BOLD-CVR, the average CVR within the region with paradoxical BOLD-CVR, PW-MRI, TCD, and stroke severity scores were obtained using the Spearman correlation coefficient with adjustment for stroke volume and the presence of a fetal posterior communicating artery.

# 3 Results

**3.1 Demographics**

We studied 20 patients with symptomatic unilateral steno-occlusive disease with clear regions of paradoxical BOLD-CVR in the ipsilateral symptomatic hemisphere (three females) with an age range of 47–86 years. Baseline characteristics are given in Table 1. Regarding neurological status, the average National Institute of Health Stroke Scale (NIHSS) score and time of admission of all patients was 4(4) and a modified Rankin scale (mRS) score of 3(2). Overall neurological improvement was observed 3 months later (NIHSS of 2(2) and mRS score of 1(1)).

**3.2 Hemodynamic Assessment of the Ipsi- and Contralateral Hemispheres**

An image of an illustrative patient is presented in Fig. 1. Average quantitative hemodynamic parameters can be found in Table 2. Only CBF did not show a significant difference between the ipsi- and contralateral regions. As shown in Fig. 2, the acute/subacute stroke patients showed the representative corresponding CBF time evolution. Chronic patients (>1 month after stroke) showed an overall clear decrease in CBF in the ipsilateral hemisphere (i.e., negative CBF difference). In TCD, the diastolic and systolic PCA-P2 flow velocities on the ipsilateral hemisphere were significantly prolonged.

**3.3 Volume and Intraregional CVR Correlation**

The volume of the region of paradoxical BOLD-CVR showed a strong negative correlation with the extent of negative BOLD-CVR values within the region of the paradoxical BOLD-CVR (rho = −0.72, $p < 0.001$). Furthermore, a high positive correlation with the ipsilateral PCA-P2 systolic flow velocity (rho = 0.59, $p = 0.005$) and diastolic flow velocity (rho = 0.70, $p = <0.001$) was found, meaning that an increasing volume was seen with more negative BOLD-CVR values, as well as increasing PCA-P2 flow

**Table 1**
**Baseline characteristics ($n = 20$)**

| | | |
|---|---|---|
| Diagnosis: number (%) | Stroke | 19 (95) |
| | TIA | 1 (5) |
| Location of affected artery: number (%) | ICA | 15 (75) |
| | MCA | 4 (20) |
| | ICA/MCA | 1 (5) |
| Lateralization: left (%) | 8 (40) | |
| Age | 66.6 ± 10.7 | |
| Sex: female (%) | 3 (15) | |
| Stroke volume ($cm^3$) | 6.7 (9.2) | |
| Presence of fetal Pcom: yes (%) | 3 (15) | |
| Vascular risk factors: yes (%) | Hypertension | 17 (85) |
| | Hypercholesterolemia | 10 (50) |
| | Smoking | 15 (75) |
| $CO_2$ during baseline | 37.5 (3.1) | |
| $CO_2$ during hypercapnia | 46.2 (3.5) | |
| $CO_2$ step change | 8.7 (2.0) | |
| BOLD-CVR over the whole brain | 0.08 ± 0.08 | |
| BOLD-CVR within the affected hemisphere | 0.06 ± 0.10 | |
| BOLD-CVR within the unaffected hemisphere | 0.12 ± 0.08 | |

*BOLD-CVR* blood oxygenation level-dependent cerebrovascular reactivity (defined as %BOLD signal change/mmHg $CO_2$), $CO_2$ carbon dioxide in mmHg

values. Moderate positive correlations were seen between the volume of the region of the paradoxical BOLD-CVR and the CBV within the ipsilateral region of negative BOLD (rho = 0.54, $p$ = 0.02) and the CVR within the contralateral region (rho = −0.53, $p$ = 0.02). No relationship was seen between the volume of the region of paradoxical BOLD-CVR and the other PW-MRI- and TCD-derived parameters.

The extent of negative BOLD-CVR values within the region of the paradoxical BOLD-CVR also showed a high negative correlation with the ipsilateral systolic PCA-P2 flow velocity (rho = −0.61, $p$ = 0.003) and diastolic flow velocity (rho = −0.59, $p$ = 0.005). Moderate negative correlations were seen with the ipsilateral CBV (rho = −0.43, $p$ = 0.03), whereas low positive correlations were found for the ipsilateral PCA-P1 flow (rho = 0.49, $p$ = 0.02) and the contralateral PCA-P2 flow (rho = −0.45, $p$ = 0.04).

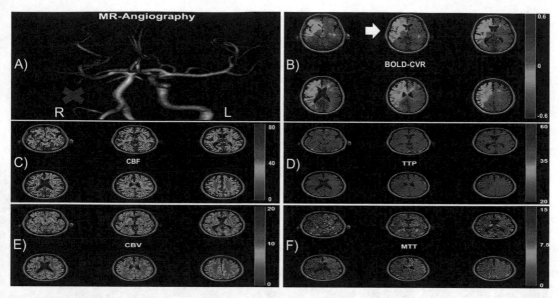

**Fig. 1** Illustrative images of a 50-year-old patient with diffuse ischemia in the right ACA/MCA territory caused by a right ICA occlusion (Panel **a**), scanned 3 weeks after stroke. Paradoxical BOLD-CVR (Panel **b**—white arrow—blue) can be observed in the right hemisphere (top right image). In the corroborating PW-MRI images (middle images), clear prolonged TTP (Panel **d**) and MTT (Panel **f**) can be seen in the region corresponding to the area of paradoxical BOLD-CVR. CBF shows a slight increase compared to the contralateral hemisphere (Panel **c**). In the CBV images, an increase in CBV within the region of paradoxical BOLD-CVR can be observed, mostly in the gray matter (Panel **e**). *BOLD-CVR* blood oxygenation level-dependent cerebrovascular reactivity (%BOLD fMRI signal change per mmHg), *CBF* cerebral blood flow, *CBV* cerebral blood volume, *ICA* internal carotid artery, *MTT* mean transit time, *TTP* time-to-peak

## 4 Discussion

In this study, we show that in patients with symptomatic unilateral steno-occlusive disease, the brain area exhibiting paradoxical BOLD-CVR in the ipsilateral symptomatic hemisphere shows evident signs of hemodynamic impairment (i.e., prolonged MTT and TTP as well as increased CBV) [1, 17, 18]. Moreover, ipsilateral to the region with paradoxical BOLD-CVR, the PCA-P2 segment flow velocity was significantly increased, and a strong positive correlation was observed between brain regions with paradoxical BOLD-CVR and ipsilateral P2 flow velocity, indicating an increased need, potentially even exhausted need, for collateralization over the posterior circulation [8]. This further supports the premise that paradoxical BOLD-CVR within the symptomatic hemisphere is a true surrogate for severe hemodynamic impairment.

### 4.1 Imaging Paradoxical BOLD-CVR

Cerebrovascular autoregulation describes the mechanism regulating the vascular tone to maintain sufficient brain tissue perfusion [19–21]. The dynamic range of this mechanism is limited, and

**Table 2**
**Quantitative hemodynamic parameters**

|  | Paradoxical BOLD-CVR region | Contralateral region | $p$-value |
|---|---|---|---|
| CVR | −0.08 (0.05) | 0.08 (0.06) | **<0.001** |
| CBF | 30.11 (9.22) | 29.45 (8.77) | 0.157 |
| CBV | 3.86 (0.96) | 3.32 (0.76) | **0.006** |
| MTT | 7.78 (1.68) | 6.74 (1.28) | **<0.001** |
| TTP | 41.47 (4.84) | 39.25 (4.08) | **<0.001** |
| **TCD** | **Ipsilateral hemisphere** | **Contralateral hemisphere** |  |
| Systolic PCA-P1 flow velocity | 88.3 (42) | 74.3 (32.8) | 0.150 |
| Diastolic PCA-P1 flow velocity | 38.8 (22.7) | 30.6 (15.1) | 0.098 |
| Systolic PCA-P2 flow velocity | 86.8 (41.7) | 60.2 (24.1) | **0.001** |
| Diastolic PCA-P2 flow velocity | 38.5 (26.4) | 24.3 (9.9) | **0.002** |

All values are reported as median (interquartile range). The contralateral region describes the region of interest on the contralateral hemisphere, determined by flipping the region with paradoxical BOLD-CVR contralaterally $P$-value after Wilcoxon rank sum test
*CBF* cerebral blood flow, *CBV* cerebral blood volume, *CVR* cerebrovascular reactivity, *MTT* mean transit time, *PCA-P1* P1 segment of the posterior cerebral artery, *PCA-P2* P2 segment of the posterior cerebral artery, *TCD* transcranial Doppler
Significance for "bold" p-value is <0.05

investigating the remaining cerebrovascular reserve capacity is therefore essential. Measurement of the perfusion reserve is widely accepted as a surrogate marker for the remaining autoregulatory blood flow control [22, 23].

An important imaging finding during perfusion reserve measurements in patients with cerebrovascular steno-occlusive disease is regions with hemodynamic failure stage 2, which can be identified as paradoxical—*exhausted*—perfusion reserve. Patients exhibiting hemodynamic failure stage 2 are at a significantly higher risk of cerebral ischemia and worse neurological outcome [22, 24–26]. Here, "exhausted perfusion reserve" describes a negative CBF response to a vasoactive stimulus. Interestingly, increased PCA-P2 segment flow has also been correlated to a higher risk of recurrent stroke, which may be due to its relationship to hemodynamic failure stage 2 [8].

Different imaging modalities, such as positron emission tomography (PET), arterial spin labelling, and single-photon emission computed tomography, can identify hemodynamic failure

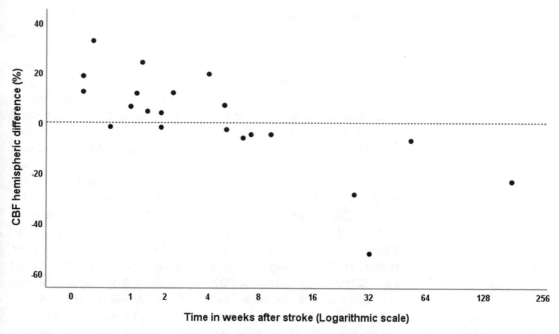

**Fig. 2** The *y*-axis describes cerebral blood flow difference in percentage between the region of paradoxical BOLD-CVR and the same region in the contralateral hemisphere as measured with perfusion-weighted MR imaging. This is set against a function of time after stroke on a logarithmic scale. A negative percent difference indicates an ipsilateral decrease in cerebral blood flow. Here, one can see the time evolution of cerebral blood flow in the region with paradoxical BOLD-CVR following stroke with increased (luxury) perfusion in some subjects during the first month as a result of inappropriate dilatation after reperfusion. *BOLD-CVR* blood oxygenation level-dependent cerebrovascular reactivity (%BOLD fMRI signal change per mmHg), *CBF* cerebral blood flow

stage 2 [2, 27, 28] However, their implementation remains challenging because of the routine clinical availability, use of radioactive isotopes, limited spatial resolution, and limited reproducibility.

To overcome these issues, BOLD-CVR has been suggested as a surrogate imaging method to measure hemodynamic failure stage 2 [2, 9, 29]. BOLD-CVR encompasses a high spatial and temporal resolution combined with a high clinical accessibility and can produce quantitative values with high inter- and intrasubject reproducibility [23, 30]. Although measurements of paradoxical BOLD-CVR have been correlated with vascular disease severity [31, 32], the pathophysiology of paradoxical BOLD-CVR remains a topic of debate [9, 10]. For instance, BOLD-CVR- and PET-derived perfusion reserve showed a good hemispheric comparison, which was in agreement with earlier findings using arterial spin labelling [2, 29, 33]. Others have detected conflicting results, thereby challenging the vascular basis of paradoxical BOLD-CVR. For instance, Arteaga et al. investigated paradoxical BOLD-CVR in 15 subjects (four with Moyamoya and 11 with unilateral or bilateral extra- or intracranial stenosis/occlusion) [10]. They found a decrease in perfusion reserve in regions with paradoxical BOLD-CVR in only three participants, whereas a significant perfusion reserve increase was seen in six subjects. The researchers hypothesized that paradoxical BOLD-CVR is not always caused by exhausted perfusion reserve but also by changes in either metabolism or CBV. Regarding the mentioned paper by Arteaga et al. [10], the premise of metabolic upregulation in the presence of an increase in cerebral blood flow is a very exciting topic but currently heavenly debated. The BOLD fMRI signal predominantly represents the relative amount of deoxyhemoglobin within a voxel. Therefore, changes in CBF, CBV, oxygen extraction fraction, cerebral metabolic rate of oxygen, and the basal amount of deoxyhemoglobin can all influence the BOLD signal response [34, 35]. There are some theories discussing metabolic influences on the BOLD signal in relation to impaired or exhausted BOLD-CVR. In this context, we have recently shown the importance of basal metabolism on the BOLD signal response to $CO_2$ [15, 36]. Papers using hypercapnia showed conflicting results regarding metabolic upregulation [35]. A current paper by Jussen et al. hints toward this possibility after showing that revascularization can restore a decrease in metabolism [37]. However, more research is necessary to investigate this potentially exciting new pathway. During this study, oxygen was kept constant during the entire protocol, and only a short hypercapnic stimulus was induced. Therefore, it can be expected that the metabolism will not or only minimally alter, and their influence is most likely negligible.

### 4.2 Limitations

These patients represent a clear subpopulation of patients with strokes. Fourteen out of 20 had an occlusion of the internal carotid artery. As a consequence, the population represents more the hemodynamic changes (chronic) due to the occlusion rather than the acute ischemic event. The chronic occurrence of the carotid pathology is also suggested by the low median NIHSS (4).

Because of the semiquantitative nature of PW-MRI, we have investigated only the region of negative BOLD-CVR within the symptomatic hemisphere and refrained from specifically examining contralateral paradoxical BOLD-CVR. Further investigations are necessary to determine the clinical value of contralateral paradoxical BOLD-CVR. The main problem of further investigations lies in the lack of quantitative CBF measurements. In contrast to BOLD-CVR, PET without the arterial input function or arterial spin labelling can be trusted only on relative measurements (between hemispheres) with a high inter- and intrasubject range, and comparisons between measurements are only reliable after normalization with the cerebellum [2, 38, 39]. Additionally, adding a metabolic measurement to determine the cerebral metabolic rate of oxygen and oxygen extraction fraction would be very beneficial in our understanding of negative BOLD-CVR and the influences of metabolic components on the BOLD signal, preferable before and after a vasoactive stimulus. We included a mixed cohort of patients with unilateral symptomatic stroke caused by cerebrovascular steno-occlusive disease of the anterior circulation. Therefore, the flow velocities within arteries of the anterior circulation as measured with TCD fluctuate much more, and we have chosen to exclude them from further evaluation. A study design investigating a more homogeneous cohort including, for instance, only subjects with internal carotid occlusions would be more appropriate to answer such a question.

## 5 Conclusion

In patients with unilateral symptomatic steno-occlusive disease, brain regions exhibiting paradoxical BOLD-CVR in the ipsilateral symptomatic hemisphere show clear signs of hemodynamic impairment, i.e., prolonged mean transit time and time-to-peak and increased cerebral blood volume. Moreover, increased flow velocities in the ipsilateral PCA-P2 segment indicate an increased need for collateralization from the posterior circulation. This further supports the premise that paradoxical BOLD-CVR within the symptomatic hemisphere is a reliable surrogate for exhausted perfusion reserve and should be studied further to determine its value for predicting stroke risk.

## Acknowledgments

This research was supported by the Forschungskredit, Postdoc Initiative 2016, from the University of Zurich (FK-16–040) and the Swiss Cancer League (KFS-3975-08-2016-R), both allocated to Dr. Jorn Fierstra.

## References

1. Esposito G, Amin-Hanjani S, Regli L (2016) Role of and indications for bypass surgery after carotid occlusion surgery study (coss)? Stroke 47:282–290

2. Fierstra J, van Niftrik C, Warnock G, Wegener S, Piccirelli M, Pangalu A et al (2018) Staging hemodynamic failure with blood oxygen-level–dependent functional magnetic resonance imaging cerebrovascular reactivity. Stroke 49:621

3. Derdeyn CP, Grubb RL Jr, Powers WJ (1999) Cerebral hemodynamic impairment: methods of measurement and association with stroke risk. Neurology 53:251–259

4. Powers WJ, Press GA, Grubb RL Jr, Gado M, Raichle ME (1987) The effect of hemodynamically significant carotid artery disease on the hemodynamic status of the cerebral circulation. Ann Intern Med 106:27–34

5. Powers WJ (2008) Imaging preventable infarction in patients with acute ischemic stroke. AJNR. Am J Neuroradiol 29:1823–1825

6. Derdeyn CP, Videen TO, Fritsch SM, Carpenter DA, Grubb RL Jr, Powers WJ (1999) Compensatory mechanisms for chronic cerebral hypoperfusion in patients with carotid occlusion. Stroke 30:1019–1024

7. Derdeyn CP, Shaibani A, Moran CJ, Cross DT III, Grubb RL Jr, Powers WJ (1999) Lack of correlation between pattern of collateralization and misery perfusion in patients with carotid occlusion. Stroke 30:1025–1032

8. Schneider J, Sick B, Luft AR, Wegener S (2015) Ultrasound and clinical predictors of recurrent ischemia in symptomatic internal carotid artery occlusion. Stroke 46:3274–3276

9. Sobczyk O, Battisti-Charbonney A, Fierstra J, Mandell DM, Poublanc J, Crawley AP et al (2014) A conceptual model for co(2)-induced redistribution of cerebral blood flow with experimental confirmation using bold mri. NeuroImage 92:56–68

10. Arteaga DF, Strother MK, Faraco CC, Jordan LC, Ladner TR, Dethrage LM et al (2014) The vascular steal phenomenon is an incomplete contributor to negative cerebrovascular reactivity in patients with symptomatic intracranial stenosis. J Cereb Blood Flow Metab 34:1453–1462

11. Slessarev M, Han J, Mardimae A, Prisman E, Preiss D, Volgyesi G et al (2007) Prospective targeting and control of end-tidal co2 and o2 concentrations. J Physiol 581:1207–1219

12. van Niftrik CHB, Piccirelli M, Bozinov O, Maldaner N, Strittmatter C, Pangalu A et al (2018) Impact of baseline co2 on blood-oxygenation-level-dependent mri measurements of cerebrovascular reactivity and task-evoked signal activation. Magn Reson Imaging 49:123

13. van Niftrik CHB, Piccirelli M, Bozinov O, Pangalu A, Fisher JA, Valavanis A et al (2017) Iterative analysis of cerebrovascular reactivity dynamic response by temporal decomposition. Brain Behav 7:e00705

14. Sebok M, van Niftrik CHB, Piccirelli M, Bozinov O, Wegener S, Esposito G et al (2018) Bold cerebrovascular reactivity as a novel marker for crossed cerebellar diaschisis. Neurology 91:e1328

15. van Niftrik CHB, Sebok M, Muscas G, Piccirelli M, Serra C, Krayenbuhl N et al (2020) Characterizing ipsilateral thalamic diaschisis in symptomatic cerebrovascular steno-occlusive patients. J Cereb Blood Flow Metab 40:563

16. Sobczyk O, Crawley AP, Poublanc J, Sam K, Mandell DM, Mikulis DJ et al (2016) Identifying significant changes in cerebrovascular reactivity to carbon dioxide. AJNR Am J Neuroradiol 37:818–824

17. Powers WJ, Zazulia AR (2010) Pet in cerebrovascular disease. PET Clin 5:83106

18. Derdeyn CP, Videen TO, Yundt KD, Fritsch SM, Carpenter DA, Grubb RL et al (2002) Variability of cerebral blood volume and oxygen extraction: stages of cerebral haemodynamic impairment revisited. Brain 125:595–607

19. Rapela CE, Green HD (1964) Autoregulation of canine cerebral blood flow. Circ Res 15 (Suppl):205–212

20. Hill MA, Davis MJ, Meininger GA, Potocnik SJ, Murphy TV (2006) Arteriolar myogenic signalling mechanisms: implications for local vascular function. Clin Hemorheol Microcirc 34:67–79

21. Lucas SJ, Tzeng YC, Galvin SD, Thomas KN, Ogoh S, Ainslie PN (2010) Influence of changes in blood pressure on cerebral perfusion and oxygenation. Hypertension 55:698–705

22. Yonas H, Smith HA, Durham SR, Pentheny SL, Johnson DW (1993) Increased stroke risk predicted by compromised cerebral blood flow reactivity. J Neurosurg 79:483–489

23. Fisher JA, Venkatraghavan L, Mikulis DJ (2018) Magnetic resonance imaging-based cerebrovascular reactivity and hemodynamic reserve: a review of method optimization and data interpretation. Stroke 49:2011

24. Webster MW, Makaroun MS, Steed DL, Smith HA, Johnson DW, Yonas H (1995) Compromised cerebral blood flow reactivity is a predictor of stroke in patients with symptomatic carotid artery occlusive disease. J Vasc Surg 21:338–344. discussion 344–335

25. Grubb RL Jr, Derdeyn CP, Fritsch SM, Carpenter DA, Yundt KD, Videen TO et al (1998) Importance of hemodynamic factors in the prognosis of symptomatic carotid occlusion. JAMA 280:1055–1060

26. Derdeyn CP, Powers WJ, Grubb RL Jr (1998) Hemodynamic effects of middle cerebral artery stenosis and occlusion. AJNR Am J Neuroradiol 19:1463–1469

27. Kuhn FP, Warnock G, Schweingruber T, Sommerauer M, Buck A, Khan N (2015) Quantitative h[o]-pet in pediatric moyamoya disease: evaluating perfusion before and after cerebral revascularization. J Stroke Cerebrovasc Dis 24:965–971

28. Derdeyn CP, Yundt KD, Videen TO, Carpenter DA, Grubb RL Jr, Powers WJ (1998) Increased oxygen extraction fraction is associated with prior ischemic events in patients with carotid occlusion. Stroke 29:754–758

29. Mandell DM, Han JS, Poublanc J, Crawley AP, Stainsby JA, Fisher JA et al (2008) Mapping cerebrovascular reactivity using blood oxygen level-dependent MRI in patients with arterial

steno-occlusive disease: comparison with arterial spin labeling MRI. Stroke 39:2021–2028

30. Kassner A, Winter JD, Poublanc J, Mikulis DJ, Crawley AP (2010) Blood-oxygen level dependent mri measures of cerebrovascular reactivity using a controlled respiratory challenge: reproducibility and gender differences. J Magn Reson Imaging 31:298–304

31. Heyn C, Poublanc J, Crawley A, Mandell D, Han JS, Tymianski M et al (2010) Quantification of cerebrovascular reactivity by blood oxygen level-dependent mr imaging and correlation with conventional angiography in patients with moyamoya disease. AJNR Am J Neuroradiol 31:862–867

32. Watchmaker JM, Frederick BD, Fusco MR, Davis LT, Juttukonda MR (2018) Lants SK, et al. Clinical use of cerebrovascular compliance imaging to evaluate revascularization in patients with moyamoya. Neurosurgery 84:261

33. Hauser TK, Seeger A, Bender B, Klose U, Thurow J, Ernemann U et al (2019) Hypercapnic bold mri compared to h2(15)o pet/ct for the hemodynamic evaluation of patients with moyamoya disease. NeuroImage Clin 22:101713

34. Davis TL, Kwong KK, Weisskoff RM, Rosen BR (1998) Calibrated functional MRI: mapping the dynamics of oxidative metabolism. Proc Natl Acad Sci U S A 95:1834–1839

35. Bright MG, Croal PL, Blockley NP, Bulte DP (2019) Multiparametric measurement of cerebral physiology using calibrated fmri. NeuroImage 187:128–144

36. Sebök M, Van Niftrik B, Piccirelli M, Bozinov O, Wegener S, Esposito G et al (2018) Bold cerebrovascular reactivity as a novel marker for crossed cerebellar diaschisis. Neurology 91:1–10

37. Jussen D, Zdunczyk A, Schmidt S, Rösler J, Buchert R, Julkunen P et al (2016) Motor plasticity after extra–intracranial bypass surgery in occlusive cerebrovascular disease. Neurology 87(1):27–35

38. Jiang TT, Videen TO, Grubb RL Jr, Powers WJ, Derdeyn CP (2010) Cerebellum as the normal reference for the detection of increased cerebral oxygen extraction. J Cereb Blood Flow Metab 30:1767–1776

39. Goetti R, Warnock G, Kuhn FP, Guggenberger R, O'Gorman R, Buck A et al (2014) Quantitative cerebral perfusion imaging in children and young adults with moyamoya disease: comparison of arterial spin-labeling-mri and h(2)[(15)o]-pet. AJNR Am J Neuroradiol 35:1022–1028

# Chapter 7

# Cerebrovascular Reactivity (CVR) in Aging, Cognitive Impairment, and Dementia

## Hanzhang Lu, Binu P. Thomas, and Peiying Liu

## Abstract

This chapter reviews current literature of CVR in the context of aging, cognitive impairment, and dementia. CVR decreases with age in a spatiotemporally specific manner, with longitudinal studies revealing more rapid rate of decline compared to cross-sectional investigations. CVR is further reduced in cognitive impairment and dementia, the mechanism of which is independent of Alzheimer's pathology. Diminishment of CVR was related to classic vascular hallmarks such as white matter hyperintensities, but with a larger spatial scope. The reduction of CVR was also found to have a negative impact on cognitive function. CVR appears to be a modifiable marker which may be useful in detecting the effect of treatment/intervention on brain vascular function. CVR may also play an important role in the interpretation of fMRI data in aging and dementia by separating vascular from neural contributions in activation signal. Existing evidences suggest that CVR may be a promising biomarker in understanding brain aging and in the diagnosis and treatment monitoring of neurodegenerative diseases.

**Key words** Cerebrovascular reactivity, Aging, Dementia, Cognitive impairment, MRI, Vascular dementia

## 1 Introduction

Dementia affects ~50 million individuals worldwide [1]. Vascular dementia (VD) [1, 2] is the second leading cause of dementia. VD and its mixed presentation with other pathologies (e.g., Alzheimer's, Parkinson's, Lewy body disease) account for more than 50% of individuals with dementia [3–6]. However, compared to other dementia types (e.g., amyloid and tau imaging in Alzheimer's disease [7–9]), validated imaging biomarkers for VD are limited. Furthermore, aging is a major risk factor for many diseases, including VD. Therefore, it is of great significance to identify biomarkers that can provide early indications of microvascular function in aging, cognitive impairment, and dementia.

The brain represents about 2% of the total body weight but consumes about 20% of the total energy [10]. Not surprisingly,

Jean Chen and Jorn Fierstra (eds.), *Cerebrovascular Reactivity: Methodological Advances and Clinical Applications*, Neuromethods, vol. 175, https://doi.org/10.1007/978-1-0716-1763-2_7, © Springer Science+Business Media, LLC, part of Springer Nature 2022

sufficient and carefully regulated blood supply is critical to meet this high energy demand and for the brain to function properly. Furthermore, brain is a spatially heterogeneous and temporally dynamic organ [11–13], with different regions requiring different amount of blood supply at different time. Therefore, the ability of the blood vessels to dilate or constrict, known as cerebrovascular reactivity (CVR), represents an important domain of vascular function and is expected to be more specific than structural changes such as white matter hyperintensities (WMH). A loss or diminishment of CVR is expected to negatively impact neural function and cognition in the context of aging and dementia.

Moreover, blood-oxygenation-level-dependent (BOLD) fMRI has been widely used to study brain functional alterations in aging and dementia. However, the BOLD fMRI signal is based on hemodynamic responses secondary to neurometabolic activations, and thus, is modulated by vascular function. Therefore, the knowledge of CVR is important for proper interpretation of BOLD fMRI signals and may provide an approach to normalize or calibrate the current fMRI technologies.

This chapter aims to describe current evidence and literature on CVR alterations in aging and dementia and how they inform vascular and neural changes under these conditions.

## 2    Cellular and Molecular Mechanisms of CVR in Relationship to Aging and Dementia

Vasodilation during a hypercapnia challenge is mediated by the relaxation of vascular smooth muscle cells (vSMCs) in the arteries and arterioles. Increased $CO_2$ content of the interstitial compartment and in the endothelial cells results in a decrease in pH through the formation of carbonic acid, followed by its dissociation into proton ($H^+$) and bicarbonate ions ($HCO_3^-$) [14]. Both the increased $CO_2$ and decreased pH can lead to SMC relaxation via the following cellular mechanisms. First, $CO_2$ and pH have a direct effect on vSMC. Specifically, increased interstitial $CO_2$ and decreased interstitial pH open potassium ($K^+$) channels on the SMCs [15, 16] and cause hyperpolarization of the vSMCs. SMC hyperpolarization decreases the activity of voltage-dependent calcium ($Ca^{2+}$) channels, resulting in reduced intracellular Ca which leads to vasodilation. Second, $CO_2$ and pH can also affect endothelium cells which indirectly dilate SMCs. Increased $CO_2$ and decreased pH can cause hyperpolarization of the endothelial cells through K channels. The negative charges of the endothelium cells can be transported to the vSMCs via myo-endothelial gap junctions, which alter the membrane potentials of the vSMCs and lead to vasodilation [17]. In addition, $CO_2$ and pH can activate nitric oxide synthase (NOS) in the endothelia [18–20], which increases the concentration of nitric oxide (NO) in the endothelia. NO then diffuses into the vSMCs where it increases cyclic GMP formation through the activation of cyclic guanylate cyclase (cGC). Activation of this pathway results in the dilation of vascular smooth muscle and results in the relaxation of vSMCs [21, 22].

Arterial aging induces vSMC phenotypic changes that lead to vascular degeneration and extracellular matrix degradation, which results in alterations of the mechanical properties of the vascular wall such as arterial stiffness [23, 24]. The relaxation of vSMCs in response to $CO_2$ and pH is also altered by aging-associated changes in several ways. Decreased expression of voltage- and $Ca^{2+}$-activated $K^+$ channels was found in some models of aging [25, 26], which directly impairs the response of vSMC relaxation to $CO_2$ and pH changes. Several studies have also demonstrated progressive decline of endothelium-dependent vasodilation with age in human [27, 28], supported by findings in animals that endothelium-dependent vasodilator responses decrease with age independently of structural changes of the vascular wall [29–31]. It has been suggested that the decline of endothelium-dependent vasodilation is primarily due to decreased NO production and/or impaired downstream signaling in the NO pathway [32]. Other risk factors associated with aging and dementia, such as hypertension [33], hypercholesterolemia [34], obesity, and diabetes [35, 36], can also contribute to small and large vessel damage and impaired vSMC relaxation. For example, it has been demonstrated in humans and animal models that hypertension alters the intracellular free $Ca^{2+}$ concentration, which leads to impaired vasorelaxation, with sustained contraction and altered regulatory myosin activity [33, 37].

Therefore, cellular and molecular changes of cerebrovascular system in aging and diseases support a hypothesis of CVR diminishment under these conditions.

## 3 Characteristics of CVR Alternations in Aging

A number of cross-sectional studies have investigated age-related difference in BOLD CVR, and most have observed that CVR decreases with age [38–40], suggesting a reduction in cerebrovascular reserve in older individuals. In terms of the magnitude of the reduction, there are some heterogeneities in the literature but are generally within a similar range. For example, Gauthier et al. observed a 26% reduction in whole-brain BOLD CVR, when comparing an older (mean age 64 years) to a younger group (mean age 24 years) of participants [39]. De Vis et al. found a 17% reduction in whole-brain BOLD CVR when comparing older (mean age 66 years) to younger group (mean age 28 years) [40]. In a lifespan study of 151 adults [38], whole-brain BOLD CVR was found to

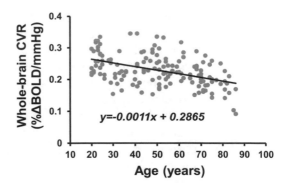

**Fig. 1** Cross-sectional relationship between whole-brain CVR (in %BOLD signal change/mmHg) and age across the adult lifespan. Each symbol represents data from one subject

decrease from 0.265%/mmHg at an age of 20 years to 0.199%/ mmHg at an age of 80 years, a reduction by 25% (Fig. 1).

Because of the complexity in BOLD signal mechanism, CVR measurements using other hemodynamic contrasts have also been performed in the context of aging. Taneja et al. compared CVR measures using phase-contrast flow MRI and arterial spin-labeling (ASL) cerebral blood flow (CBF) MRI [41]. Comparing 17 older (mean age 62 years) to 32 younger participants (mean age 22 years), it was observed that phase-contrast MRI was more sensitive in detecting age differences (a 17% between younger and older group), presumably because of the relatively high signal-to-noise ratio (SNR) and straightforward quantification of this technique [41]. The limitation of phase-contrast MRI is that it provides a whole-brain measure only, but does not have spatial information. ASL-based CVR, on the other hand, did not reveal a significant difference between the groups ($p = 0.31$), although the numerical CVR value was lower in older than in younger group (by 9%) [41]. These findings suggest that ASL CVR, while having the potential advantages of quantitative interpretation and spatial resolution, is less suited for the study of CVR in the context of aging, presumably because of the technical issues associated with this method such as relatively lower SNR and susceptibility to other physiological parameters such as bolus arrival time and labeling efficiency. An absence of age effect in ASL CVR was also noted by De Vis et al. [40].

Longitudinal studies of CVR were performed less frequently but have also been conducted. In a 4-year longitudinal study, Peng et al. showed that there was a longitudinal decrease in CVR between two time points and the rate of reduction (−0.0024%/ mmHg/year) was considerably higher than that observed with cross-sectional data (−0.0011%/mmHg/year) [42]. These findings suggest that cross-sectional observations may underestimate age-related changes, presumably because of sampling biases in that older participants enrolled in research studies may represent a super-normal group of individuals. Another advantage of longitudinal studies is that it allows the calculation of CVR decline rate on an individual basis, which revealed that middle-aged, but not advanced age, individuals showed the fastest rate of CVR decline across the lifespan [42]. This observation is consistent with reports that vascular risk factors during middle age can predict cognitive decline and amyloid accumulation some 20 years later [43].

## 4  Spatial Features of CVR Change in Aging

CVR is symmetric across hemispheres and homogeneously distributed in gray matter (GM) in healthy young volunteers [44]. CVR data from a lifespan cohort revealed a significant reduction in CVR with age across most of the brain, major lobes, and subcortical structures except the occipital pole [38]. Another study by Peng et al. longitudinally followed 116 participants and observed spatial heterogeneity in CVR decline in aging, in that temporal lobe showed the fastest rate of CVR decline, followed by parietal and frontal lobes, followed by occipital lobe. In another study, CVR was significantly lower in the frontal, temporal, and occipital lobe of older compared to young subjects [40]. This study, however, did not observe difference in CVR in deep GM structures and the parietal lobe [40]. A study by Leoni et al. assessed CVR using BOLD and ASL MRI [45]. They observed that ASL-based CVR decreased in the cerebral cortex and subcortical regions in older subjects, while BOLD-based CVR amplitude was lower and time-to-peak (TTP) was greater in GM and in posterior regions in the brain in older compared to young subjects [45]. BOLD-based CVR findings were confirmed in the study by Thomas et al. [46]. They observed that cortical GM CVR had smaller amplitude and greater TTP (i.e., it took longer to observe the CVR response in GM) in the older compared to younger subjects [46]. The opposite was observed in the innermost core of the white matter (WM), with larger CVR amplitude and smaller TTP (i.e., CVR response was observed earlier in WM) in the older compared to younger subjects, which was attributed to WM tissue being less densely packed due to age-related changes, making it easier for blood to flow through WM [46]. Reich et al. determined CVR by measuring CBF using Xenon-133 and reported decreased cortical GM CVR in aging [47]. Another study assessed CVR to changing end-tidal $CO_2$ using a step and ramp $CO_2$ stimulus delivery. In response to the ramp stimulus, CVR amplitude in the elderly subjects was reduced in the frontal WM but was similar in the GM when compared to the young subjects, whereas the step stimulus did not yield any differences [48].

Age-related changes in CVR have revealed significant heterogeneity in signal amplitude and TTP in different regions in the brain and provide substantial knowledge on changes in the dynamic vascular properties throughout the brain accompanied with aging.

## 5 CVR Changes in Cognitive Impairment and Dementia

Vascular disease represents an important contribution to cognitive impairment and dementia [1, 2]. It is known that structural damage, typically evaluated by WMH on T2-weighted magnetic resonance imaging, is a hallmark of small vessel disease. However, WMH represents a late event and only accounts for a small fraction, if any, variance in cognition [49–51]. There has been growing evidence showing that CVR is a more sensitive early biomarker in cognitive impairment and dementia.

Although earlier studies using the gamma camera ($Xe^{133}$) [52, 53], SPECT ($^{99m}Tc$-hexamethylpropyleneamineoxime) [54, 55], and PET ($^{15}O$) showed mixed results [56, 57], it has been well-established by transcranial Doppler ultrasonography studies that CVR of the middle cerebral artery is reduced in mild cognitive impairment (MCI) and dementia patients [58–61]. Studies of MRI-based CVR mapping in MCI and AD patients are limited but emerging. Using 3 T MRI with hypercapnia challenge (5% $CO_2$ inhalation), Yezhuvath et al. compared CVR in 17 early AD and 17 normal participants and found that sporadic AD patients have diminished CVR compared to elderly controls [62]. The CVR deficit region in the AD group was mainly in frontal lobe and insula. In a study using 1.5 T MRI and carbogen inhalation, Cantin et al. examined CVR in nine AD, seven MCI, and 11 normal controls and found decreased CVR in both AD and MCI groups compared to normal controls, with no significant difference between MCI and AD groups [63]. In contrast to the Yezhuvath et al.'s finding [62], the CVR reduction in AD group was found to be dominant in posterior areas (parietal and occipital cortex, posterior cingulate) [63]. This regional discrepancy could be due to technical and sampling differences between the two studies. In a recent study with larger sample size by Sur et al. [64], which included 43 cognitive impaired individuals and 26 healthy elderly, whole-brain CVR was found to be significantly lower in the impaired group.

Effects of other risk factors of AD and vascular pathology on CVR have also been reported. Although no effect of apoE4 genotype was detected by Cantin et al. [63] probably due to limited sample size, widespread CVR deficits have been demonstrated among APOE ?4 gene carriers in both older [65] and young adults [66] and most notably in the prefrontal and parahippocampal regions. In older adults, the cognitive effect of APOE ε4 was found to be magnified in hypertension and low $CO_2$ individuals [65]. Interestingly, in a longitudinal study by Peng et al. [42], where 116 healthy adults across the lifespan were studied 4 years apart, individuals with hypertension revealed a lower CVR compared to normotensive participants cross-sectionally, but the rate of longitudinal CVR change was not different between hypertensive and normotensive participants [42].

## 6 The Relationship of CVR with Cognitive Function and Other Clinical Measures

CVR has been shown to affect cognition in aging and dementia populations, although the exact cognitive domain(s) that is most affected is still not fully elucidated. In an impaired cohort with mixed contributions of Alzheimer's and vascular pathology, Sur et al. found that absolute CVR (in %/mmHg) of the whole brain was significantly associated with global cognitive function indexed by Montreal Cognitive Assessment (MoCA) or by a composite neuropsychological battery score [64]. These associations were independent of amyloid and tau concentrations measured from cerebrospinal fluid (CSF), suggesting that CVR affects cognition through a pathway or mechanism separate from the Alzheimer's pathology. In terms of individual cognitive domains, the authors found that whole-brain absolute CVR was associated with language score, but not with episodic memory, executive function, or processing speed, although all coefficients were positive in sign. Regional investigations revealed that CVR across different brain lobes showed similar relationship with cognition as the whole-brain value. Interestingly, in a separate study in patients with dementia, CVR was found to be correlated to the score of Boston Naming Testing, again suggesting that CVR is particularly associated with language function in impaired and demented older individuals.

In non-impaired aging populations, longitudinal studies revealed that 4-year changes in whole-brain CVR were associated with changes in cognitive functions in the domains of episodic memory and processing speed [42]. Note, however, that language was not specifically assessed in that study. In an aging study of calibrated fMRI, Hutchison et al. showed that the ratio between task-related CBF change and CMRO2 change, an index equivalent to CVR, was associated with response time during a visual luminance detection task, suggesting that CVR is related to processing speed [67]. Collectively, questions remain as to which cognitive domain is most susceptible to impairment in CVR. It is also possible that cognitive domains are differentially affected in non-impaired and impaired older populations. From the current literature, it appears that regional, e.g., lobar, CVR provides generally similar information as whole-brain CVR in terms of predicting cognitive performance in aging and dementia studies.

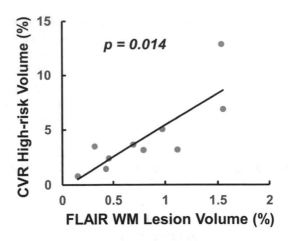

**Fig. 2** The relationship between volume of white matter hyperintensities (WMH) and volume of CVR high-risk tissues. CVR high-risk tissues are defined as brain voxels with CVR values lower than 10% of whole-brain CVR. Each symbol represents data from one subject

CVR was found to be related to white matter hyperintensities (WMH) in a number of manners. CVR in WMH was found to be lower than normal-appearing white matter, consistent with a vascular ischemic hypothesis for WMH [68]. Whole-brain WMH volume was not related to whole-brain CVR. However, WMH volume was related to the volume of "CVR high-risk voxels" (Fig. 2), which are defined as voxels with CVR values less than 10% of global CVR [62]. These high-risk regions are typically five times larger than the volume of WMH, suggesting that much of the normal-appearing white matter in an individual with considerable WMH already showed signs of early vascular malfunction. This notion was also supported by observations that CVR in normal-appearing white matter of individuals with significant WMH was lower than that in individuals without WMH [68]. In a study by Sam et al., CVR in WMH areas was found to be significantly lower than contralateral normal-appearing WM (NAWM) in the same subjects [69]. A 1-year longitudinal follow-up study in older subjects reported that areas of NAWM that develop WM hyperintensities over time have reduced CVR compared to areas of NAWM that do not [70]. Recently studies also found that CVR in stroke lesion regions was lower than the contralateral normal regions [71].

# 7    Intervention and Treatment Effects on CVR

There are some evidences suggesting that CVR is a modifiable marker. Aerobic exercise has been shown to improve CVR. For example, Murrell et al. reported that a 12-week aerobic exercise intervention increased CVR as measured by the CBF velocity in the middle cerebral artery (MCA) using Doppler ultrasound in young and aging subjects in response to $CO_2$ inhalation [72]. This result was confirmed by another cross-sectional study which reported that CVR as measured by CBF velocity using Doppler ultrasound was higher in physically active and fit elderly women compared to older women in response to hypercapnia, again confirming the benefits of physical activity and exercise on CVR [73]. Bailey et al. assessed the effects of lifelong aerobic exercise on CVR in a cross-sectional study [74]. They reported that the highly fit elderly that have exercised for around 45 years of their lives had higher CVR as measured by CBF velocity in the MCA compared to sedentary elderly [74]. These findings confirm that aerobic exercise can help prevent age-related decline in CVR and highlight the importance of maintaining aerobic fitness throughout the lifespan given its capacity to improve CVR later in life. Another study performed a 7-month aerobic exercise intervention in elderly and reported that CBF velocity increased in both left and right MCA in response to breath-holding, which is also known to increase the pressure of $CO_2$ in the lungs [75]. Not all exercise intervention studies have reported increase in CVR. A study by Chapman et al. reported no difference in CVR measured by BOLD MRI in a group that performed aerobic exercise for 12 weeks compared to another group that performed cognitive training for the same duration [76]. Neither exercise nor cognitive training for 12 weeks was shown to increase CVR [76]. There have also been reports of decrease in CVR with exercise. A study by Thomas et al. reported that CVR measured by BOLD MRI was paradoxically lower in the whole brain, major lobes, cerebellum, and subcortical structures in elderly master athletes who performed aerobic exercises for most of their life compared to sedentary elderly [77]. Skeletal muscles of the elderly athletes produce $CO_2$ due to the constant exercise they perform, which crosses the blood-brain barrier. They interpreted the reduced CVR as brain vasculature being desensitized to the constant exposure to $CO_2$ and suggested it being protection mechanism to prevent brain vasculature from over-dilating during exercise when demand for blood is more in the limbs [77]. These findings were later confirmed by two other groups. DuBose et al. reported lower CVR measured by BOLD MRI after 3-month-long chronic aerobic exercise training [78]. Intzandt et al. also reported lower CVR as measured by BOLD and ASL MRI with increase aerobic fitness in elderly subjects [79]. Drugs have also shown to have effects on CVR. Study by Kastrup et al. reported that hormone replacement therapy in postmenopausal women was shown to enhance CVR [80]. Another study reported impairment of CVR by methionine-induced hyper-homocysteinemia and amelioration by quinapril treatment [81]. Finally, a study showed that, after a single dose of sildenafil administration in Alzheimer's patients, CVR in response to $CO_2$ inhalation was decreased which was accompanied by an elevation in basal CBF, suggesting that sildena-fil increased basal vascular tone but lowered the vascular reserve [82].

## 8    Application of CVR in the Normalization of fMRI Signal

Cognitive aging and dementia studies using BOLD fMRI have revealed a wealth of information about the aging brain. For example, in the aging literature, despite the declines of cognitive performance and brain volumetrics with age [83–85], older adults often show heightened frontal activation relative to young [86–89], which is thought to represent a compensatory activation that occurs in the aging brain to accommodate the decreased volume of neural tissue and declining efficiency of neural circuitry [84, 90]. In contrast to the increased activation of the frontal regions, age-related fMRI signal decrease is typically represented in posterior brain regions such as visual areas [88, 91–94] and medial temporal lobe [86, 87, 89], which is interpreted as a functional decline with age. These findings have been replicated reliably by many laboratories.

However, BOLD fMRI signal is an indirect measure of neural activity [95], and it relies on the dilation of blood vessels [96]. Several studies have demonstrated that the BOLD signal amplitude is modulated by vascular reactivity [91, 97–99]. More recent studies have found that resting-state functional connectivity measured by BOLD fMRI also varies with CVR [100–102]. Therefore, an important but largely unaddressed point is that these BOLD fMRI signal patterns observed in the aging and dementia literature cannot be directly interpreted as neural activity changes given that vasodilatory capacity is known to decline with age [38, 42] and dementia [58–64].

A potential approach to account for vascular component in BOLD signal is to measure CVR and include it as a covariate or normalization factor in the data analysis. It has been suggested that accounting for CVR can reduce BOLD signal variations [96, 97] and therefore improve its sensitivity in detecting age-related neural activity changes [91, 103]. Importantly, it has also been demonstrated that accounting for the vascular declines could refine our understanding of brain functional changes measured by BOLD fMRI [91, 104–107]. For example, in a study by Liu et al., 130 subjects ranging from 20 to 89 years old underwent both a memory-encoding task fMRI and a hypercapnia CVR mapping scan [105]. The fMRI responses showed an age-related decline in the occipital and medial temporal cortices while manifesting an age-related increase in the right frontal cortex, consistent with previous literature [86–89, 91–93]. On the other hand, after accounting for vascular declines, CVR-corrected fMRI signal revealed bilateral frontal over-recruitment in older individuals, whereas none of the brain regions manifested an age-related decrease in the CVR-corrected fMRI signal [105], which is distinctive from the vast majority of existing aging literature.

Another possible approach to obtain a more accurate estimation of neural activity is to use a calibrated fMRI method [108–114]. Calibrated fMRI is based on the BOLD biophysical model, and, by experimentally estimating parameters of the model using hypercapnia and/or hyperoxia challenges, one can separate the neural effect from vascular contributions. This method has shown great promises in quantitative evaluation of the BOLD effect and has been applied in studies of aging [39, 40, 115, 116] and AD [101, 117]. However, a technical challenge of the calibrated fMRI approach is that the data acquisition requires the use of simultaneous BOLD and arterial spin-labeling (ASL) pulse sequence, which is often associated with lower signal-to-noise ratio (SNR) and reduced brain coverage.

# 9    Conclusion

In this chapter, we reviewed current literature of CVR in the context of aging, cognitive impairment, and dementia. CVR decreases with age in a spatiotemporally specific manner, with longitudinal studies revealing more rapid rate of decline compared to cross-sectional investigations. CVR is further reduced in cognitive impairment and dementia, the mechanism of which is independent of Alzheimer's pathology. Diminishment of CVR was related to classic vascular hallmarks such as white matter hyperintensities, but with a larger spatial scope. The reduction of CVR was also found to have a negative impact on cognitive function. CVR appears to be a modifiable marker which may be useful in detecting the effect of treatment/intervention on brain vascular function. CVR may also play an important role in the interpretation of fMRI data in aging and dementia by separating vascular from neural contributions in activation signal. Existing evidences suggest that CVR may be a promising biomarker in understanding brain aging and in the diagnosis and treatment monitoring of neurodegenerative diseases.

# References

1. WHO (2017) Dementia. https://www.who.int/en/news-room/fact-sheets/detail/dementia

2. Alzheimer's Association (2018) 2018 Alzheimer's disease facts and figures. Alzheimers Dement 14(3):367–429

3. Fernando MS, Ince PG, Function MRCC, Ageing Neuropathology Study G (2004) Vascular pathologies and cognition in a population-based cohort of elderly people. J Neurol Sci 226(1–2):13–17

4. Schneider JA, Arvanitakis Z, Bang W, Bennett DA (2007) Mixed brain pathologies account for most dementia cases in community-dwelling older persons. Neurology 69:2197–2204

5. Esiri MM, Nagy Z, Smith MZ, Barnetson L, Smith AD (1999) Cerebrovascular disease and threshold for dementia in the early stages of Alzheimer's disease. Lancet 354 (9182):919–920

6. Schneider JA, Wilson RS, Bienias JL, Evans DA, Bennett DA (2004) Cerebral infarctions and the likelihood of dementia from Alzheimer disease pathology. Neurology 62:1148–1155

7. Klunk WE (2011) Amyloid imaging as a biomarker for cerebral beta-amyloidosis and risk prediction for Alzheimer dementia. Neurobiol Aging 32(Suppl 1):S20–S36

8. Saint-Aubert L, Lemoine L, Chiotis K, Leuzy A, Rodriguez-Vieitez E, Nordberg A (2017) Tau PET imaging: present and future directions. Mol Neurodegener 12(1):19

9. Jack CR Jr, Wiste HJ, Weigand SD, Therneau TM, Lowe VJ, Knopman DS et al (2017) Defining imaging biomarker cut points for brain aging and Alzheimer's disease. Alzheimers Dement 13(3):205–216

10. Rolfe DFS, Brown GC (1997) Cellular energy utilization and molecular origin of standard metabolic rate in mammals. Physiol Rev 77:731–758

11. Gupta L, Besseling RM, Overvliet GM, Hofman PA, de Louw A, Vaessen MJ et al (2014) Spatial heterogeneity analysis of brain activation in fMRI. NeuroImage Clin 5:266–276

12. Fu Z, Tu Y, Di X, Biswal BB, Calhoun VD, Zhang Z (2017) Associations between functional connectivity dynamics and BOLD dynamics are heterogeneous across brain networks. Front Hum Neurosci 11:593

13. Daitch AL, Parvizi J (2018) Spatial and temporal heterogeneity of neural responses in human posteromedial cortex. Proc Natl Acad Sci U S A 115(18):4785–4790

14. Yoon SH, Zuccarello M, Rapoport RM (2000) Reversal of hypercapnia induces endothelin-dependent constriction of basilar artery in rabbits with acute metabolic alkalosis. Gen Pharmacol 35(6):333–340

15. Brayden JE (1996) Potassium channels in vascular smooth muscle. Clin Exp Pharmacol Physiol 23(12):1069–1076

16. Peng HL, Jensen PE, Nilsson H, Aalkjaer C (1998) Effect of acidosis on tension and [Ca2+]i in rat cerebral arteries: is there a role for membrane potential? Am J Phys 274(2 Pt 2):H655–H662

17. Sandow SL, Haddock RE, Hill CE, Chadha PS, Kerr PM, Welsh DG et al (2009) What's where and why at a vascular myoendothelial microdomain signalling complex. Clin Exp Pharmacol Physiol 36(1):67–76

18. Fathi AR, Yang C, Bakhtian KD, Qi M, Lonser RR, Pluta RM (2011) Carbon dioxide influence on nitric oxide production in endothelial cells and astrocytes: cellular mechanisms. Brain Res 1386:50–57

19. Xu HL, Koenig HM, Ye S, Feinstein DL, Pelligrino DA (2004) Influence of the glia limitans on pial arteriolar relaxation in the rat. Am J Physiol Heart Circ Physiol 287(1):H331–H339

20. Ziegelstein RC, Cheng L, Blank PS, Spurgeon HA, Lakatta EG, Hansford RG et al (1993) Modulation of calcium homeostasis in cultured rat aortic endothelial cells by intracellular acidification. Am J Phys 265(4 Pt 2):H1424–H1433

21. Bolotina VM, Najibi S, Palacino JJ, Pagano PJ, Cohen RA (1994) Nitric oxide directly activates calcium-dependent potassium channels in vascular smooth muscle. Nature 368 (6474):850–853

22. Zhao Y, Vanhoutte PM, Leung SW (2015) Vascular nitric oxide: beyond eNOS. J Pharmacol Sci 129(2):83–94

23. Monk BA, George SJ (2015) The effect of ageing on vascular smooth muscle cell behaviour--a mini-review. Gerontology 61 (5):416–426

24. Lacolley P, Regnault V, Avolio AP (2018) Smooth muscle cell and arterial aging: basic and clinical aspects. Cardiovasc Res 114 (4):513–528

25. Marijic J, Li Q, Song M, Nishimaru K, Stefani E, Toro L (2001) Decreased expression of voltage- and Ca(2+)-activated K(+) channels in coronary smooth muscle during aging. Circ Res 88(2):210–216

26. Albarwani S, Al-Siyabi S, Baomar H, Hassan MO (2010) Exercise training attenuates ageing-induced BKCa channel downregulation in rat coronary arteries. Exp Physiol 95 (6):746–755

27. Hatake K, Kakishita E, Wakabayashi I, Sakiyama N, Hishida S (1990) Effect of aging on endothelium-dependent vascular relaxation of isolated human basilar artery to thrombin and bradykinin. Stroke 21 (7):1039–1043

28. Egashira K, Inou T, Hirooka Y, Kai H, Sugimachi M, Suzuki S et al (1993) Effects of age on endothelium-dependent vasodilation of resistance coronary artery by acetylcholine in humans. Circulation 88(1):77–81

29. Murohara T, Yasue H, Ohgushi M, Sakaino N, Jougasaki M (1991) Age related

attenuation of the endothelium dependent relaxation to noradrenaline in isolated pig coronary arteries. Cardiovasc Res 25 (12):1002–1009

30. Mantelli L, Amerini S, Ledda F (1995) Roles of nitric oxide and endothelium-derived hyperpolarizing factor in vasorelaxant effect of acetylcholine as influenced by aging and hypertension. J Cardiovasc Pharmacol 25 (4):595–602

31. Chinellato A, Pandolfo L, Ragazzi E, Zambonin MR, Froldi G, De Biasi M et al (1991) Effect of age on rabbit aortic responses to relaxant endothelium-dependent and endothelium-independent agents. Blood Vessels 28(5):358–365

32. Sonntag WE, Eckman DM, Ingraham J, Riddle DR (2007) Chapter 12 Regulation of cerebrovascular aging. In: Riddle DR (ed) Brain aging: models, methods, and mechanisms. CRC Press/Taylor & Francis, Boca Raton, FL

33. Touyz RM, Alves-Lopes R, Rios FJ, Camargo LL, Anagnostopoulou A, Arner A et al (2018) Vascular smooth muscle contraction in hypertension. Cardiovasc Res 114(4):529–539

34. Azadzoi KM (1991) Saenz de Tejada I. Hypercholesterolemia impairs endothelium-dependent relaxation of rabbit corpus cavernosum smooth muscle. J Urol 146(1):238–240

35. Sivitz WI, Wayson SM, Bayless ML, Sinkey CA, Haynes WG (2007) Obesity impairs vascular relaxation in human subjects: hyperglycemia exaggerates adrenergic vasoconstriction arterial dysfunction in obesity and diabetes. J Diabetes Complicat 21(3):149–157

36. Doyon G, Bruemmer D (2016) Vascular smooth muscle cell dysfunction in diabetes: nuclear receptors channel to relaxation. Clin Sci 130(20):1837–1839

37. Takeya K, Wang X, Sutherland C, Kathol I, Loutzenhiser K, Loutzenhiser RD et al (2014) Involvement of myosin regulatory light chain diphosphorylation in sustained vasoconstriction under pathophysiological conditions. J Smooth Muscle Res 50:18–28

38. Lu H, Xu F, Rodrigue KM, Kennedy KM, Cheng Y, Flicker B et al (2011) Alterations in cerebral metabolic rate and blood supply across the adult lifespan. Cereb Cortex 21 (6):1426–1434

39. Gauthier CJ, Madjar C, Desjardins-Crepeau-L, Bellec P, Bherer L, Hoge RD (2013) Age dependence of hemodynamic response characteristics in human functional magnetic resonance imaging. Neurobiol Aging 34 (5):1469–1485

40. De Vis JB, Hendrikse J, Bhogal A, Adams A, Kappelle LJ, Petersen ET (2015) Age-related changes in brain hemodynamics; A calibrated MRI study. Hum Brain Mapp 36 (10):3973–3987

41. Taneja K, Liu P, Xu C, Turner M, Zhao Y, Abdelkarim D et al (2020) Quantitative cerebrovascular reactivity in normal aging: comparison between phase-contrast and arterial spin labeling MRI. Front Neurol 11:758

42. Peng SL, Chen X, Li Y, Rodrigue KM, Park DC, Lu H (2018) Age-related changes in cerebrovascular reactivity and their relationship to cognition: a four-year longitudinal study. NeuroImage 174:257–262

43. Gottesman RF, Albert MS, Alonso A, Coker LH, Coresh J, Davis SM et al (2017) Associations between midlife vascular risk factors and 25-year incident dementia in the atherosclerosis risk in communities (ARIC) cohort. JAMA Neurol 74(10):1246–1254

44. van der Zande FH, Hofman PA, Backes WH (2005) Mapping hypercapnia-induced cerebrovascular reactivity using BOLD MRI. Neuroradiology 47(2):114–120

45. Leoni RF, Oliveira IA, Pontes-Neto OM, Santos AC, Leite JP (2017) Cerebral blood flow and vasoreactivity in aging: an arterial spin labeling study. Braz J Med Biol Res 50 (4):e5670

46. Thomas BP, Liu P, Park DC, van Osch MJ, Lu H (2014) Cerebrovascular reactivity in the brain white matter: magnitude, temporal characteristics, and age effects. J Cereb Blood Flow Metab 34(2):242–247

47. Reich T, Rusinek H (1989) Cerebral cortical and white matter reactivity to carbon dioxide. Stroke 20(4):453–457

48. McKetton L, Sobczyk O, Duffin J, Poublanc J, Sam K, Crawley AP et al (2018) The aging brain and cerebrovascular reactivity. NeuroImage 181:132–141

49. He J, Wong VS, Fletcher E, Maillard P, Lee DY, Iosif AM et al (2012) The contributions of MRI-based measures of gray matter, white matter hyperintensity, and white matter integrity to late-life cognition. AJNR Am J Neuroradiol 33(9):1797–1803

50. Knopman DS, Griswold ME, Lirette ST, Gottesman RF, Kantarci K, Sharrett AR et al (2015) Vascular imaging abnormalities and cognition: mediation by cortical volume in nondemented individuals: atherosclerosis risk in communities-neurocognitive study. Stroke 46(2):433–440

51. Warren MW, Weiner MF, Rossetti HC, McColl R, Peshock R, King KS (2015) Cognitive impact of lacunar infarcts and white matter hyperintensity volume. Demen Geriatr Cogn Disord Extra 5(1):170–175

52. Hachinski VC, Iliff LD, Zilhka E, Du Boulay GH, McAllister VL, Marshall J et al (1975) Cerebral blood flow in dementia. Arch Neurol 32(9):632–637

53. Yamaguchi F, Meyer JS, Yamamoto M, Sakai F, Shaw T (1980) Noninvasive regional cerebral blood flow measurements in dementia. Arch Neurol 37(7):410–418

54. Stoppe G, Schutze R, Kogler A, Staedt J, Munz DL, Emrich D et al (1995) Cerebrovascular reactivity to acetazolamide in (senile) dementia of Alzheimer's type: relationship to disease severity. Dementia 6(2):73–82

55. Pavics L, Grunwald F, Reichmann K, Sera T, Ambrus E, Horn R et al (1998) rCBF SPECT and the acetazolamide test in the evaluation of dementia. Nucl Med Rev C E Eur 1(1):13–19

56. Kuwabara Y, Ichiya Y, Otsuka M, Masuda K, Ichimiya A, Fujishima M (1992) Cerebrovascular responsiveness to hypercapnia in Alzheimer's dementia and vascular dementia of the Binswanger type. Stroke 23(4):594–598

57. Jagust WJ, Eberling JL, Reed BR, Mathis CA, Budinger TF (1997) Clinical studies of cerebral blood flow in Alzheimer's disease. Ann N Y Acad Sci 826:254–262

58. Vicenzini E, Ricciardi MC, Altieri M, Puccinelli F, Bonaffini N, Di Piero V et al (2007) Cerebrovascular reactivity in degenerative and vascular dementia: a transcranial Doppler study. Eur Neurol 58(2):84–89

59. Viticchi G, Falsetti L, Vernieri F, Altamura C, Bartolini M, Luzzi S et al (2012) Vascular predictors of cognitive decline in patients with mild cognitive impairment. Neurobiol Aging 33(6):1127.e1–1127.e9

60. Lee ST, Jung KH, Lee YS (2007) Decreased vasomotor reactivity in Alzheimer's disease. J Clin Neurol 3(1):18–23

61. Bar KJ, Boettger MK, Seidler N, Mentzel HJ, Terborg C, Sauer H (2007) Influence of galantamine on vasomotor reactivity in Alzheimer's disease and vascular dementia due to cerebral microangiopathy. Stroke 38(12):3186–3192

62. Yezhuvath US, Uh J, Cheng Y, Martin-Cook K, Weiner M, Diaz-Arrastia R et al (2012) Forebrain-dominant deficit in cerebrovascular reactivity in Alzheimer's disease. Neurobiol Aging 33(1):75–82

63. Cantin S, Villien M, Moreaud O, Tropres I, Keignart S, Chipon E et al (2011) Impaired cerebral vasoreactivity to CO2 in Alzheimer's disease using BOLD fMRI. NeuroImage 58(2):579–587

64. Sur S, Lin Z, Li Y, Yasar S, Rosenberg P, Moghekar A et al (2020) Association of cerebrovascular reactivity and Alzheimer pathologic markers with cognitive performance. Neurology 95(8):e962–ee72

65. Hajjar I, Sorond F, Lipsitz LA (2015) Apolipoprotein E, carbon dioxide vasoreactivity, and cognition in older adults: effect of hypertension. J Am Geriatr Soc 63(2):276–281

66. Suri S, Mackay CE, Kelly ME, Germuska M, Tunbridge EM, Frisoni GB et al (2015) Reduced cerebrovascular reactivity in young adults carrying the APOE epsilon4 allele. Alzheimers Dement 11(6):648–57.e1

67. Hutchison JL, Lu H, Rypma B (2013) Neural mechanisms of age-related slowing: the DeltaCBF/DeltaCMRO2 ratio mediates age-differences in BOLD signal and human performance. Cereb Cortex 23(10):2337–2346

68. Uh J, Yezhuvath U, Cheng Y, Lu H (2010) In vivo vascular hallmarks of diffuse leukoaraiosis. J Magn Reson Imaging 32(1):184–190

69. Sam K, Crawley AP, Poublanc J, Conklin J, Sobczyk O, Mandell DM et al (2016) Vascular dysfunction in leukoaraiosis. AJNR Am J Neuroradiol 37(12):2258–2264

70. Sam K, Crawley AP, Conklin J, Poublanc J, Sobczyk O, Mandell DM et al (2016) Development of white matter hyperintensity is preceded by reduced cerebrovascular reactivity. Ann Neurol 80(2):277–285

71. Taneja K, Lu H, Welch BG, Thomas BP, Pinho M, Lin D et al (2019) Evaluation of cerebrovascular reserve in patients with cerebrovascular diseases using resting-state MRI: a feasibility study. Magn Reson Imaging 59:46–52

72. Murrell CJ, Cotter JD, Thomas KN, Lucas SJ, Williams MJ, Ainslie PN (2013) Cerebral blood flow and cerebrovascular reactivity at rest and during sub-maximal exercise: effect of age and 12-week exercise training. Age (Dordr) 35(3):905–920

73. Brown AD, McMorris CA, Longman RS, Leigh R, Hill MD, Friedenreich CM et al (2010) Effects of cardiorespiratory fitness and cerebral blood flow on cognitive outcomes in older women. Neurobiol Aging 31(12):2047–2057

74. Bailey DM, Marley CJ, Brugniaux JV, Hodson D, New KJ, Ogoh S et al (2013) Elevated aerobic fitness sustained throughout

the adult lifespan is associated with improved cerebral hemodynamics. Stroke 44 (11):3235–3238

75. Vicente-Campos D, Mora J, Castro-Pinero J, Gonzalez-Montesinos JL, Conde-Caveda J, Chicharro JL (2012) Impact of a physical activity program on cerebral vasoreactivity in sedentary elderly people. J Sports Med Phys Fitn 52(5):537–544

76. Chapman SB, Aslan S, Spence JS, Keebler MW, DeFina LF, Didehbani N et al (2016) Distinct brain and behavioral benefits from cognitive vs. physical training: a randomized trial in aging adults. Front Hum Neurosci 10:338

77. Thomas BP, Yezhuvath US, Tseng BY, Liu P, Levine BD, Zhang R et al (2013) Life-long aerobic exercise preserved baseline cerebral blood flow but reduced vascular reactivity to CO2. J Magn Reson Imaging 38 (5):1177–1183

78. Dubose LE, Wharff C, Schmid P, Muellerleile M, Sigurdsson G, Reist L et al (2019) Chronic aerobic exercise training reduces cerebrovascular reactivity to a breath hold stimulus in middle-aged and older adults. Exp Biol 33:lb431

79. Intzandt B, Sabra D, Foster C, Desjardins-Crepeau L, Hoge RD, Steele CJ et al (2020) Higher cardiovascular fitness level is associated with lower cerebrovascular reactivity and perfusion in healthy older adults. J Cereb Blood Flow Metab 40(7):1468–1481

80. Kastrup A, Dichgans J, Niemeier M, Schabet M (1998) Changes of cerebrovascular CO2 reactivity during normal aging. Stroke 29 (7):1311–1314

81. Chao CL, Lee YT (2000) Impairment of cerebrovascular reactivity by methionine-induced hyperhomocysteinemia and amelioration by quinapril treatment. Stroke 31 (12):2907–2911

82. Sheng M, Lu H, Liu P, Li Y, Ravi H, Peng SL et al (2017) Sildenafil improves vascular and metabolic function in patients with Alzheimer's disease. J Alzheimers Dis 60 (4):1351–1364

83. Raz N, Lindenberger U, Rodrigue KM, Kennedy KM, Head D, Williamson A et al (2005) Regional brain changes in aging healthy adults: general trends, individual differences and modifiers. Cereb Cortex 15 (11):1676–1689

84. Park DC, Reuter-Lorenz P (2009) The adaptive brain: aging and neurocognitive scaffolding. Annu Rev Psychol 60:173–196

85. Bartzokis G, Beckson M, Lu PH, Nuechterlein KH, Edwards N, Mintz J (2001) Age-related changes in frontal and temporal lobe volumes in men: a magnetic resonance imaging study. Arch Gen Psychiatry 58 (5):461–465

86. Park DC, Welsh RC, Marshuetz C, Gutchess AH, Mikels J, Polk TA et al (2003) Working memory for complex scenes: age differences in frontal and hippocampal activations. J Cogn Neurosci 15(8):1122–1134

87. Gutchess AH, Welsh RC, Hedden T, Bangert A, Minear M, Liu LL et al (2005) Aging and the neural correlates of successful picture encoding: frontal activations compensate for decreased medial-temporal activity. J Cogn Neurosci 17(1):84–96

88. Cabeza R, Daselaar SM, Dolcos F, Prince SE, Budde M, Nyberg L (2004) Task-independent and task-specific age effects on brain activity during working memory, visual attention and episodic retrieval. Cereb Cortex 14(4):364–375

89. Daselaar SM, Veltman DJ, Rombouts SA, Raaijmakers JG, Jonker C (2003) Neuroanatomical correlates of episodic encoding and retrieval in young and elderly subjects. Brain 126(Pt 1):43–56

90. Cabeza R, Anderson ND, Locantore JK, McIntosh AR (2002) Aging gracefully: compensatory brain activity in high-performing older adults. NeuroImage 17(3):1394–1402

91. Handwerker DA, Gazzaley A, Inglis BA, D'Esposito M (2007) Reducing vascular variability of fMRI data across aging populations using a breathholding task. Hum Brain Mapp 28(9):846–859

92. Ross MH, Yurgelun-Todd DA, Renshaw PF, Maas LC, Mendelson JH, Mello NK et al (1997) Age-related reduction in functional MRI response to photic stimulation. Neurology 48(1):173–176

93. Buckner RL, Snyder AZ, Sanders AL, Raichle ME, Morris JC (2000) Functional brain imaging of young, nondemented, and demented older adults. J Cogn Neurosci 12 (Suppl 2):24–34

94. Davis SW, Dennis NA, Daselaar SM, Fleck MS, Cabeza R (2008) Que PASA? The posterior-anterior shift in aging. Cereb Cortex 18(5):1201–1209

95. Ogawa S, Menon RS, Tank DW, Kim SG, Merkle H, Ellermann JM et al (1993) Functional brain mapping by blood oxygenation level-dependent contrast magnetic resonance imaging. A comparison of signal

characteristics with a biophysical model. Biophys J 64(3):803–812

96. Bandettini PA, Wong EC (1997) A hypercapnia-based normalization method for improved spatial localization of human brain activation with fMRI. NMR Biomed 10 (4–5):197–203

97. Thomason ME, Foland LC, Glover GH (2007) Calibration of BOLD fMRI using breath holding reduces group variance during a cognitive task. Hum Brain Mapp 28 (1):59–68

98. Liau J, Liu TT (2009) Inter-subject variability in hypercapnic normalization of the BOLD fMRI response. NeuroImage 45(2):420–430

99. Liu P, Hebrank AC, Rodrigue KM, Kennedy KM, Park DC, Lu H (2013) A comparison of physiologic modulators of fMRI signals. Hum Brain Mapp 34(9):2078–2088

100. Golestani AM, Kwinta JB, Strother SC, Khatamian YB, Chen JJ (2016) The association between cerebrovascular reactivity and resting-state fMRI functional connectivity in healthy adults: the influence of basal carbon dioxide. NeuroImage 132:301–313

101. Lajoie I, Nugent S, Debacker C, Dyson K, Tancredi FB, Badhwar A et al (2017) Application of calibrated fMRI in Alzheimer's disease. NeuroImage Clin 15:348–358

102. Chu PPW, Golestani AM, Kwinta JB, Khatamian YB, Chen JJ (2018) Characterizing the modulation of resting-state fMRI metrics by baseline physiology. NeuroImage 173:72–87

103. Song Z, McDonough IM, Liu P, Lu H, Park DC (2016) Cortical amyloid burden and age moderate hippocampal activity in cognitively-normal adults. NeuroImage Clin 12:78–84

104. D'Esposito M, Zarahn E, Aguirre GK, Rypma B (1999) The effect of normal aging on the coupling of neural activity to the bold hemodynamic response. NeuroImage 10(1):6–14

105. Liu P, Hebrank AC, Rodrigue KM, Kennedy KM, Section J, Park DC et al (2013) Age-related differences in memory-encoding fMRI responses after accounting for decline in vascular reactivity. NeuroImage 78:415–425

106. Riecker A, Grodd W, Klose U, Schulz JB, Groschel K, Erb M et al (2003) Relation between regional functional MRI activation and vascular reactivity to carbon dioxide during normal aging. J Cereb Blood Flow Metab 23(5):565–573

107. Kannurpatti SS, Motes MA, Rypma B, Biswal BB (2011) Increasing measurement accuracy of age-related BOLD signal change: minimizing vascular contributions by resting-state-fluctuation-of-amplitude scaling. Hum Brain Mapp 32(7):1125–1140

108. Kim SG, Ugurbil K (1997) Comparison of blood oxygenation and cerebral blood flow effects in fMRI: estimation of relative oxygen consumption change. Magn Reson Med 38 (1):59–65

109. Davis TL, Kwong KK, Weisskoff RM, Rosen BR (1998) Calibrated functional MRI: mapping the dynamics of oxidative metabolism. Proc Natl Acad Sci U S A 95 (4):1834–1839

110. Hoge RD, Atkinson J, Gill B, Crelier GR, Marrett S, Pike GB (1999) Linear coupling between cerebral blood flow and oxygen consumption in activated human cortex. Proc Natl Acad Sci U S A 96(16):9403–9408

111. Chiarelli PA, Bulte DP, Piechnik S, Jezzard P (2007) Sources of systematic bias in hypercapnia-calibrated functional MRI estimation of oxygen metabolism. NeuroImage 34(1):35–43

112. Chiarelli PA, Bulte DP, Wise R, Gallichan D, Jezzard P (2007) A calibration method for quantitative BOLD fMRI based on hyperoxia. NeuroImage 37(3):808–820

113. Uludag K, Dubowitz DJ, Yoder EJ, Restom K, Liu TT, Buxton RB (2004) Coupling of cerebral blood flow and oxygen consumption during physiological activation and deactivation measured with fMRI. NeuroImage 23(1):148–155

114. Sicard KM, Duong TQ (2005) Effects of hypoxia, hyperoxia, and hypercapnia on baseline and stimulus-evoked BOLD, CBF, and CMRO2 in spontaneously breathing animals. NeuroImage 25(3):850–858

115. Ances BM, Liang CL, Leontiev O, Perthen JE, Fleisher AS, Lansing AE et al (2009) Effects of aging on cerebral blood flow, oxygen metabolism, and blood oxygenation level dependent responses to visual stimulation. Hum Brain Mapp 30(4):1120–1132

116. Mohtasib RS, Lumley G, Goodwin JA, Emsley HC, Sluming V, Parkes LM (2012) Calibrated fMRI during a cognitive Stroop task reveals reduced metabolic response with increasing age. NeuroImage 59 (2):1143–1151

117. Fleisher AS, Podraza KM, Bangen KJ, Taylor C, Sherzai A, Sidhar K et al (2009) Cerebral perfusion and oxygenation differences in Alzheimer's disease risk. Neurobiol Aging 30(11):1737–1748

# Chapter 8

# Magnetic Resonance Imaging Methods for Assessment of Hemodynamic Reserve in Chronic Steno-occlusive Cerebrovascular Disease

Keith R. Thulborn, Laura Stone McGuire, Fady T. Charbel, and Sepideh Amin-Hanjani

## Abstract

Assessment of cerebrovascular hemodynamic reserve is an important component of evaluating patients with steno-occlusive cerebrovascular disease, as loss of reserve can be indicative of elevated future stroke risk. Here we describe multimodal magnetic resonance imaging (MRI) techniques for assessment of cerebrovascular reserve, including (1) local blood oxygenation level-dependent (BOLD) signal arising from increased regional blood flow with task performance during functional MRI, using a task paradigm that interrogates cortical areas in all vascular territories simultaneously, (2) global BOLD signal response occurring as a result of increased widespread cerebral blood flow from hypercapnia as a vasodilatory challenge, and (3) direct large cerebral vessel volumetric flow measurements using phase-contrast quantitative MR angiography to assess flow before and after global vasodilation with intravenous acetazolamide. Absence of the expected responses allows detection of compromised or absent hemodynamic reserve. The techniques for performing the different MRI modalities are discussed including relative advantages and limitations.

**Key words** Cerebral blood flow, Cerebral ischemia, Echo planar imaging, Functional MRI, Hemodynamic reserve, Intracranial arteriosclerosis, Magnetic resonance imaging (MRI), Magnetic resonance angiography (MRA), Moyamoya disease

## 1 Introduction

This chapter summarizes our last decade of clinical experience using quantitative flow measurements from phase-contrast MR angiography (QMRA) and functional magnetic resonance (MR) imaging (fMRI) in its various forms for the management of chronic cerebrovascular disease. These MR imaging approaches provide complementary information for the assessment of cerebral hemodynamic reserve (HR) beyond that of catheter-based cerebral angiography, single-photon emission computed tomography (SPECT), and computed tomography (CT) perfusion and angiography. Acute stroke protocols are not discussed, as MR imaging

Jean Chen and Jorn Fierstra (eds.), *Cerebrovascular Reactivity: Methodological Advances and Clinical Applications*, Neuromethods, vol. 175, https://doi.org/10.1007/978-1-0716-1763-2_8, © Springer Science+Business Media, LLC, part of Springer Nature 2022

plays only a secondary role at our institution, given the guidelines for immediate cerebral endovascular thrombectomy and use of thrombolytic agents. Clinical management of chronic cerebrovascular disease has multiple possible interventions with different degrees of invasiveness that need to be tailored to the predicted risk of future stroke in each patient. Hemodynamic reserve (HR), the capacity of cerebral tissue to regulate regional blood flow to match regional metabolic requirements despite changes in systemic pressures, can be used as one means to assess and monitor risk of future stroke. Methods of assessing HR, using QMRA and functional MR imaging, suitable for a clinical service, are described in detail and illustrated with specific cases. The chapter is presented in three sections for clarity. Subheading 2 is a comprehensive review of the role of QMRA for flow measurements in the large intracranial and extracranial arteries for the management of chronic cerebrovascular disease. Subheading 3 is a detailed discussion of the clinical methodology of blood oxygenation level-dependent (BOLD) fMRI using a task-based paradigm and a hypercapnia challenge paradigm that can be used practically on a clinical service. Subheading 4 serves to illustrate the application of the QMRA and fMRI methods in the management of several clinical cases. These cases are examples of moyamoya disease and atherosclerotic disease from our clinical service. The cited references include both the historical development of the concepts and the latest insights in management of chronic cerebrovascular disease.

## 2  Section 1

### 2.1  Overview

Loss of hemodynamic reserve (HR) in patients with chronic steno-occlusive cerebrovascular disease can be indicative of elevated stroke risk. Quantitative flow measurements in large cerebral vessels before and after global vasodilation with intravenous acetazolamide using phase-contrast quantitative MR angiography (PC MRA) provides an estimation of the integrity of HR in the corresponding distal vascular territories. Absence of the expected increase in blood flow reflects compromised or absent hemodynamic reserve in that distal tissue. Our experience over more than a decade with using quantitative MR angiography (QMRA) based on a commercially available software package on a clinical MR imaging service will be described. The QMRA methodology is discussed emphasizing quality assurance in Subheading 2 and illustrated through the clinical case examples in Subheading 4. These cases integrate QMRA and fMRI modalities to ascertain the integrity of HR through these independent complementary approaches.

### 2.2  Introduction to Quantitative MRA

QMRA uses both time-of-flight (TOF) and phase-contrast (PC) techniques to acquire vascular images [1–5]. The TOF method provides high-resolution vascular anatomy, either in 3D or 2D modes, and the PC method offers a noninvasive measure of vascular velocities and vessel cross-sectional areas to estimate the flow rate (mL/min) through specified vessels. While the TOF technique offers structural insight into the disease pathology and may be useful for screening purposes (i.e., detection of a stenotic artery), the quantitative flows from the PC technique characterize the hemodynamic implications of the disease and allow monitoring of disease progression, particularly with the addition of a vasodilatory challenge.

These techniques use endogenous contrast arising from blood flow to depict vessel lumens and characterize blood flow. TOF MRA visualizes a vessel lumen due to flow-related signal enhancement of fully relaxed longitudinal magnetization of water protons in fresh blood entering into the imaging slice against a background of saturated longitudinal magnetization of stationary water protons in the surrounding tissue. Signal contrast of the vessel is reduced under conditions of increased time in the imaging plane that results in increasing saturation of longitudinal magnetization. This may occur when the vessel is flowing within the imaging plane, when turbulent flow increases time within the imaging plane, or due to slow flow. PC MRA uses the phase shift that occurs when magnetized blood flows through an applied magnetic field gradient. The phase shift of the magnetization moving in the same direction as the applied magnetic field gradient is proportional to the flow velocity of the magnetization. PC MRA determines the velocity and cross-sectional area of the vessel from which the blood flow can be calculated in quantitative units of milliliter per minute.

PC MRA quantification of blood flow has been studied extensively and validated in multiple settings. In vitro studies with phantom models have shown minimal inaccuracy rates ranging from 4% to 10% in both steady and pulsatile flow conditions [6–11]. In vivo studies have similarly validated flow measurements in a variety of systems, from larger vessels such as the aorta [9, 12–15] to smaller vessels [16, 17], including all cerebral arteries [18–25]. Typically, PC MRA measures arterial flow at different time points during the cardiac cycle termed "cine mode." This may be performed with cardiac gating [20, 22, 26] or without cardiac gating [19, 27–30].

QMRA combines both techniques using TOF MRA to visualize the anatomy of blood vessels and PC MRA to measure blood flow. The specifics of the QMRA protocol as established by Zhao et al. are detailed below [31]. The software used for QMRA of the cerebrovascular system is currently commercially available as the NOVA (Noninvasive Optimal Vessel Analysis) system (VasSol, Inc., Chicago, IL USA).

### 2.3 Methodology of Quantitative MRA Using NOVA

#### 2.3.1 Generation of 3D Models of the Vasculature from TOF MRA

Zhao et al. [31] reported a 3D vessel localization technique to calculate the double oblique imaging plane through any arterial segment visualized on a 3D model of the vasculature obtained from the TOF MR angiogram and then used cardiac-gated PC MRA to measure the flow velocity and cross-sectional area of the selected vessel to quantify blood flow in that vessel [31].

The extracranial vessels of the neck are evaluated with 2D TOF MRA using multiple axial image slices acquired with a flow-compensated gradient-echo sequence. The intracranial vessels are evaluated with 3D multiple overlapping thin-slab TOF MRA. Both types of data are rendered into 3D models of the vascular structures using the marching cube algorithm that constructs a 3D polygonal mesh of the isosurface of the vascular structures depicted by TOF MRA [32] (Fig. 1). This algorithm has been used extensively in computer graphic design for surface rendering, particularly in medical imaging [32].

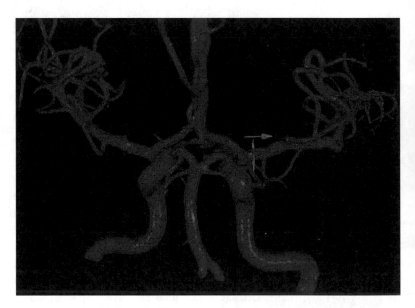

**Fig. 1** 3D model of the TOF MRA generated by the NOVA software using the surface-rendering marching cube algorithm. The software allows the model to be rotated around the three orthogonal axes to allow the user to select segments of vessels for flow measurements using PC MRA. The double oblique plane required to measure flow velocity most accurately is calculated automatically by the NOVA software once the vessel segment is selected. That information is then sent to the MR scanner for automated prescription of the imaging plane used for the cine PC measurement. The cardiac gating is performed with peripheral pulse oximetry. In this example, the left middle cerebral artery had been selected. The green arrow indicates the direction of flow. The yellow plane of PC imaging is shown through the M1 segment of the left middle cerebral artery

*2.3.2  Vessel Selection and PC MRA*

Within the NOVA software, the 3D model of the vasculature can be rotated and magnified to allow the user to select straight vessel segments from which flow measurements can be made. The software calculates the double oblique imaging planes that are perpendicular to the vessel flow axis of each of the selected vessels. This perpendicular orientation minimizes the error of the flow velocity measurement. The orientation error increases as this orthogonal imaging plane angle decreases from 90° to 0° from the flow direction. The information for the prescription of the double oblique imaging plane for each selected vessel is sent to the MR scanner. The scanner performs separate cardiac-gated acquisitions at 12 time points across the cardiac cycle in a single imaging plane tailored to each selected vessel. Those PC images are then transferred offline back to the NOVA software where a region of interest (ROI) is automatically placed around the vessel of interest on each of the cine PC images obtained through the cardiac cycle to isolate the changing phase values of all voxels within that vessel. The phase variation through a vessel cross section is converted to a discrete color-coded flow velocity map (Fig. 2). The user can refine this automated ROI, if necessary, to better define the vessel boundary threshold. The wall of the vessel is stationary ideally, and so the phase does not vary across the cardiac cycle. The blood in the middle of the lumen has the highest flow velocity and so has the highest phase variation across the cardiac cycle. For large vessel flow, the phase and therefore flow velocity profile of a vessel have parabolic profiles that give the cross-sectional appearance of concentric rings of colors (Fig. 2). Vessel profiles that are not circular have not been imaged in the orthogonal plane to flow indicating large errors may be present in measured flows. This is usually due to

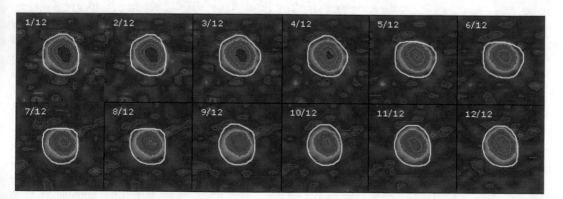

**Fig. 2** The matrix of the 12 partitions of the flow velocity cross-sectional profiles of the PC images, acquired at each of the 12 time points through the cardiac cycle, shows the variation in flow velocity across the left middle cerebral artery shown in Fig. 1. The cross section is circular indicating the appropriate plane of acquisition, and the flow velocity increases from the periphery (blue) to the center of the lumen (red during systole, partition 1/12; yellow during diastole, partition 9/12). The white line around the lumen on each partition is the automated selection of the lumen circumference. This contour can be modified by the user, if necessary

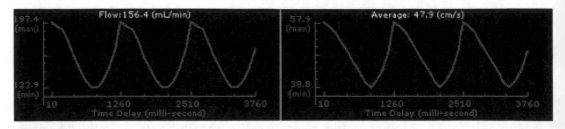

**Fig. 3** The waveform of flow (left, mL/min) and flow velocity (right, cm/s) in the left middle cerebral artery from Figs. 1 and 2, calculated from the 12 time points across the cardiac cycle by PC MRA. The highest flow and velocity are during systole, and the lowest values are during diastole. NOVA software calculates a mean flow and velocity across the cardiac cycle for the pictogram summary shown in Fig. 4 (i.e., 156 mL/min for left middle cerebral artery)

patient motion. The temporal waveforms that are generated automatically from the flow velocity variation through the cardiac cycle should also be inspected to ensure that the cardiac cycle is well represented with systolic and diastolic phases, sometimes with a dichroic notch (Fig. 3). Poor peripheral pulse oximetry used for cardiac gating can be the source of error in flow measurements. The direction of the phase encoding gradient is along the flow direction and is determined by the double oblique imaging prescription already discussed. The strength of the phase encoding gradient must also be set based on the expected flow velocity. As phase encoding has a limited range of values in which phase is unambiguous ($-90°$ to $+90°$), higher flows must use a higher gradient strength to set the maximum phase range above the highest flow expected in the vessel to avoid phase encoding aliasing. This parameter is termed the VENC or velocity encoding parameter. Low flows must use a lower VENC value to accurately encode the phase change of slow flow. This VENC setting is another source of error in measurement that must be minimized by appropriate choice for each vessel. The software estimates a high and low VENC value for different vessels and then suggests changes if the phase values indicate a need for readjustment and repeat measurements on the scanner. This VENC check by NOVA minimizes errors but consumes more time for the repeated data transfers and acquisitions.

Intrarater and interrater reliability pose little concern for producing reproducible flow measurements, with coefficients of variation for intrarater reliability of 5% for nine vessels and 5–10% for three vessels; the mean coefficients of variation for interrater reliability were less than 4% when tested on the basilar artery and left middle cerebral artery [33, 34]. The largest source of error is due to patient motion during the examination. As the TOF MRA is performed first and used for the prescription of the imaging planes for subsequent PC MRA measurements, any motion between these measurements results in large errors. This means that the studies must be done efficiently as even the most cooperative patients tend to move over time. This is especially true for the examinations performed with a vasodilatory challenge. Another source of error for comparison of flows between different examinations is head orientation that may influence phase measurements by changing eddy currents. Eddy currents are minimized on scanners along the orthogonal imaging axes but may not be fully compensated when the gradients are mixed for the double oblique prescriptions used by this software. This factor increases the variation in flow measurements across different examinations on the same patient that may influence interpretation of disease progression.

**2.3.3 Pharmacologic Vasodilatory Challenge**

Cerebrovascular reactivity refers to the normal response of cerebral capillaries and precapillary arterioles to vasoactive stimuli, such as carbon dioxide, which results in vasodilation to reduce vascular resistance with subsequent increased blood flow through the larger vessels as well as the capillaries. This cerebral autoregulation response provides adequate blood delivery to the brain based on regional metabolic demand independent of systemic pressures. However, cerebrovascular reactivity may become dysfunctional in several disease states, including steno-occlusive disease. In steno-occlusive disease, the resistance vessels distal to a stenotic artery can become habitually dilated as a compensatory mechanism to preserve blood supply [35, 36]. As these distal vessels become chronically dilated with worsened severity of the stenosis, a lack of the expected increased blood flow is seen in response to a global vasodilatory stimulus. In the most severe hemodynamically compromised tissue, the distal vasculature becomes maximally vasodilated, and a paradoxical reduction in cerebral blood flow, known as a steal phenomenon, is encountered whereby blood is preferentially routed into the surrounding healthier tissue in response to the vasodilatory challenge.

Vasodilatory challenges allow operational assessment of cerebrovascular reactivity. Several options exist, including inhaled carbon dioxide and breath-holding techniques [37, 38]. However, the acquisition time in QMRA poses challenges for routine clinical application. Intravenous acetazolamide offers a more suitable avenue for a vasodilatory challenge for clinical QMRA [39]. Acetazolamide is a carbonic anhydrase inhibitor, slowing the conversion of $CO_2$ to bicarbonate ions for transport away from the brain tissue and thereby causing the accumulation of carbon dioxide in brain tissue with resultant vasodilation. Following the baseline QMRA measurements, slow intravenous administration of acetazolamide (1 g dissolved in sterile water for adults without contraindications) over 5 min and a delay of a further 10 min achieve adequate vasodilation. A second QMRA study is then performed in the same vessels using the same prescriptions and compared to the baseline flow measurements. The flow changes induced by vasodilation and measured on QMRA can be interpreted in terms of the integrity of HR [40]. Increases in vessel flow in the range of 40% are expected; reduced or absent responses indicate impaired HR [40].

*2.3.4  Features of NOVA QMRA to Minimize Errors*

The commercially available software, Noninvasive Optimal Vessel Analysis (NOVA, VasSol, Inc., Chicago, IL, USA), uses a standardized TOF MRA and PC MRA protocol to perform QMRA. This software addresses several limitations associated with performing flow quantification using PC MRA. The automatic defining of the double oblique imaging plane to measure flow velocity perpendicular to the flow direction tailored to all selected vessels minimizes flow errors from variations in vessel orientation. Having defined the direction of the flow velocity encoding gradient, the NOVA system provides an automated selection of magnitudes for flow encoding gradient reflected in the VENC parameter. After the PC measurements are analyzed, NOVA verifies the validity of VENC selection, allowing a repeat measurement on the scanner at a more appropriate VENC value if necessary to avoid phase encoding aliasing. This is important for diseased vessels where flows may vary significantly from normal. NOVA QMRA has been validated with both in vitro phantom simulators [11, 31] and in vivo canine model [41].

*2.3.5  MR Imaging Parameters*

NOVA QMRA may be completed on either 1.5 or 3.0 T MR systems (GE Healthcare, Siemens Healthineers USA, Philips Healthcare) using standard pulse sequences. The 3D MRA TOF parameters for the head are usually Repitition Time/Echo Time (TR/TE), 23/3.3 ms; flip angle, 20; Field of View (FOV), 200 mm; section thickness, 1 mm; and matrix, 512 × 256. TOF MRA images are transferred to a separate workstation running NOVA software to complete the 3D surface-rendered vascular model described above. A user selects the vessel segments of interest from this 3D model, and the NOVA software calculates the double oblique PC imaging plane prescription parameters that are sent to the MR scanner. The scanner acquires the cine PC data for each vessel segment in each prescribed plane at the VENC settings determined by NOVA. The cine PC data are returned from the scanner to the NOVA software that then automatically derives estimates of flow in each selected vessel. The PC data obtained using the retrospectively cardiac-gated cine PC technique uses 12 time points across the cardiac cycle, and the following imaging acquisition parameters are set: TR, 10–13 ms; TE, 4–7 ms; flip angle, 15; Number of excitations (NEX), 4; slice thickness, 4 mm for basilar artery and 3 mm for other vessels; FOV, 160 mm for vertebral arteries and 140 mm for other arteries; and matrix, 256 × 192. The automated vessel diameter can be refined by the user, as necessary. The NOVA software then generates a flow report for all selected vessels tabulating all of the mean flow rates (mL/min) along with normal values for three adult age ranges (18–40, 41–60, and >60 years), a 3D model showing the location and plane of the measurement and direction of flow (Fig. 1), cross-sectional flow velocity profile matrix through the cardiac cycle (Fig. 2), and flow variation through the cardiac cycle (Fig. 3) for each selected vessel.

**2.4 Data Interpretation**

The NOVA QMRA software outputs a pictogram summarizing all of the flow values and direction of flow in each of the selected vessels as shown in Fig. 4. This report is sent to and displayed on PACS for interpretation. This pictogram provides a global view of the data, but each flow measurement is subject to important quality assurance steps during interpretation. The flow value for each vessel has the 3D model of the vasculature showing the double oblique plane of imaging through the selected vessel (Fig. 1), the corresponding vessel cross-sectional flow profile (Fig. 2), and the temporal waveform across the cardiac cycle (Fig. 3). These additional data should be inspected to ensure that the measurement is valid before clinical interpretation is made.

If interpretation is being performed on a workstation running NOVA software, the navigation tools allow the 3D model of the vasculature to be rotated freely around the three orthogonal axes. The operator may quickly identify the vessels and location of each measurement to verify that the correct vessel has been selected and that there was no head movement between the TOF MRA and the PC MRA measurements. Waveforms are used to verify that the cardiac gating was appropriate and that the flow profile of the vessel cross section is valid. Incorrect setting of VENC alters this profile. Poor cardiac gating that can cause large errors in flow can be detected. The report contains all of this information, and careful review is warranted to achieve the best interpretation.

The range of normal flow values for extracranial and intracranial vessels has been determined with a prospective study of 325 healthy adult volunteers [42]. Table 1 shows a summary of the normal value ranges for volumetric flow rate in the following vessels: common carotid artery, internal carotid artery, vertebral artery, basilar artery, middle cerebral artery, anterior cerebral artery, and posterior cerebral artery.

**2.4.1 QMRA in Assessment of Cerebral Blood Flow**

QMRA has been used to determine regional [43] and total cerebral blood flow in healthy adults [20, 29, 44, 45]. Several studies have shown a progressive decline of cerebral blood flow with age [19, 42, 46]. A study of 250 adults found a significant yearly decrease with advancing age in total cerebral blood flow (TCBF) of 4.8 mL/min and no sex differences [19]. Similarly, in a prospective cohort of 325 adult healthy participants, with ages ranging

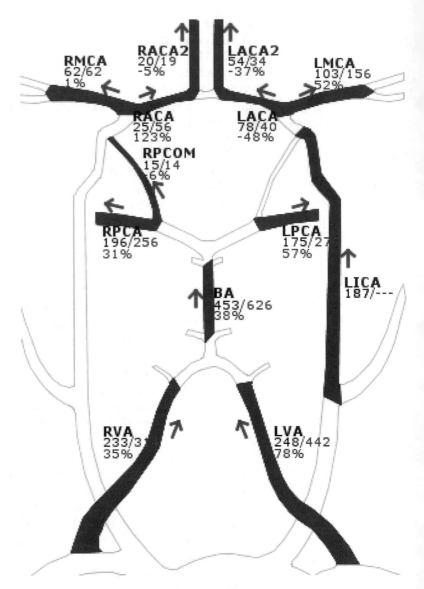

**Fig. 4** Pictogram of the circle of Willis summarizing the blood flows (mL/min) for the intracranial vessels that can be measured (marked in red) including the right (R) and left (L) internal carotid arteries (RICA, LICA), right and left vertebral arteries (RVA, LVA), basilar artery (BA), right and left posterior cerebral arteries (RPCA, LPCA), A1 segments of the right and left anterior cerebral arteries (RACA, LACA), A2 segments of the right and left anterior cerebral arteries (RACA2, LACA2), and right and left middle cerebral artery (RMCA, LMCA). Vessels in which measurements were not made are shown by the white lines without red filling. The direction of flow is indicated by green arrows. Flows are also measured at baseline (numbers in black) and after vasodilation (numbers in blue). The percentage change in flow is calculated below the flows (red numbers labeled %). The results are sent to PACS for review and interpretation

**Table 1**
**Vessel flow rates with age group differences [42]**

|       | All | 18–40 y.o. | 41–60 y.o. | 61–80 y.o. |
|-------|-----|-----------|-----------|-----------|
| TCBF  | 706 ± 111 | 749 ± 115 | 700 ± 107 | 656 ± 103 |
| LCCA  | 392 ± 72  | 422 ± 75  | 382 ± 68  | 368 ± 73  |
| RCCA  | 389 ± 75  | 407 ± 79  | 389 ± 68  | 369 ± 76  |
| LICA  | 259 ± 50  | 273 ± 46  | 255 ± 51  | 243 ± 53  |
| RICA  | 256 ± 52  | 267 ± 52  | 257 ± 49  | 235 ± 53  |
| LVA   | 100 ± 38  | 110 ± 38  | 97 ± 37   | 88 ± 37   |
| RVA   | 88 ± 30   | 89 ± 31   | 89 ± 25   | 87 ± 33   |
| BA    | 138 ± 41  | 149 ± 44  | 137 ± 41  | 124 ± 31  |
| LMCA  | 159 ± 28  | 170 ± 33  | 159 ± 27  | 146 ± 22  |
| RMCA  | 146 ± 28  | 159 ± 31  | 144 ± 24  | 133 ± 26  |
| LPCA  | 69 ± 14   | 72 ± 14   | 68 ± 12   | 61 ± 12   |
| RPCA  | 66 ± 15   | 72 ± 16   | 65 ± 14   | 58 ± 12   |
| LACA  | 86 ± 26   | 94 ± 26   | 80 ± 23   | 79 ± 26   |
| RACA  | 93 ± 28   | 96 ± 32   | 94 ± 26   | 86 ± 32   |

*A2* ACA post-communicating segment, *ACA* anterior cerebral artery, *BA* basilar artery, *CCA* common carotid artery, *ICA* internal carotid artery, *L* left, *MCA* middle cerebral artery, *n* sample size (reflects exclusion of outliers), *PCA* posterior cerebral artery, *PR* posterior region, *R* right, *T* total, *TCBF* total cerebral blood flow, *VA* vertebral artery, *y.o.* years old
Flow rates in mL/min ± 1 s.d. in the major cerebral vessels

from 18 to 84 years old, all individual vessel flows and total cerebral blood flow (TCBF) declined with age, at 2.6 mL/min per year for TCBF, but TCBF did not demonstrate any relationship relative to sex or race [42].

QMRA has also been used to characterize the distribution of CBF within the circle of Willis. In a prospective study of 208 patients, the ICA volume flow in subjects with a complete configuration of the circle of Willis was 245 mL/min, and the presence of fetal-type PCAs or the absence of A1 segments altered the ICA input [21]. A similar finding was seen in the prospective cohort of 325 participants, which found both decreased basilar artery flow and increased ICA flow in subjects with fetal PCAs and increased flow in the ICA with a hypoplastic ACA [42]. The effects of age and vascular anatomy on vessel-specific flows have important implications for interpreting the hemodynamic impact of cerebrovascular occlusive disease states.

**2.5  Clinical Relevance of QMRA in Steno-occlusive Disease**

*2.5.1  Atherosclerotic Disease*

QMRA aids in the assessment of the underlying pathophysiology in multiple cerebrovascular disorders [47], including steno-occlusive disease. Hemodynamic assessment with QMRA has been used to identify patients who may have high risk for stroke, and because recent studies have shown high 2-year ischemic stroke rates of approximately 20% in symptomatic intracranial stenosis patients [48], this imaging modality offers an important opportunity for further evaluation. A study of 33 patients with unilateral ICA occlusion found that the symptomatic group demonstrated significantly less total sum of volume flow rate in ipsilateral anterior, middle, and posterior cerebral arteries when compared to the stable or asymptomatic groups [49]. Additionally, QMRA can assess leptomeningeal collateral flow in large vessel cerebrovascular disease [50]. Assessment of leptomeningeal collaterals could provide an important marker of cerebrovascular reserve because loss of reserve in chronic steno-occlusive disease has been shown to be associated with poor leptomeningeal collaterals on digital subtraction angiography. In a study of QMRA in combination with a Diamox challenge, patients with robust leptomeningeal collaterals had significantly better flow change within the ipsilateral vascular territory of the MCA than patients with poor leptomeningeal collaterals [40]. However, direct measurement of distal collateral blood flow can prove challenging due to the small caliber of the vessels under consideration, and there remains inconsistency in grading systems [51].

Flow measurements using QMRA have been shown to be useful for assessment of subsequent stroke risk in patients with symptomatic vertebrobasilar (VB) disease [52, 53]. Patients can be designated as low flow or normal flow based on a flow algorithm which uses large vessel flow in the VB territory to evaluate distal territory regional hemodynamic status [52]. In a retrospective study of 47 patients, those with normal flow had event-free survival (stroke or any ischemic event) at 2 years of 100% and 96%, respectively, compared to those with low distal flow, with event-free survival of 71% and 53%, respectively [52]. A larger follow-up prospective cohort study of patients with atherosclerotic VB disease yielded similar results with 78% and 96% stroke-free survival at 12 months in low- and normal-flow patients, respectively. The hazard ratio, adjusted for age and other stroke risk factors, in the low distal flow status group was 11.55 [53].

QMRA offers insight into additional steno-occlusive diseases, including subclavian steal syndrome [54] and extracranial carotid stenosis [55, 56], and their effect on intracranial blood flow. A case series demonstrated findings of vertebral artery flow reversal in the setting of symptomatic subclavian steal [54]. A study of 38 patients with unilateral >65% carotid stenosis found that plaque length and vessel diameter were independent predictors of ICA flow [55]. In a prospective study of 24 patients with >60% carotid stenosis undergoing carotid endarterectomy, patients with preoperative impairment of MCA blood flow on QMRA were more likely to experience improvement in flow after revascularization, which was associated with greater cognitive improvement in attention and executive functioning [56].

Quantitative parameters allow the practitioner to monitor disease progression over time and to evaluate the impact of interventions. Current management paradigms of patients with intracranial steno-occlusive atherosclerotic disease include first lifestyle modifications and medical management with antiplatelet therapy. If aggressive medical management fails and the patient presents with persistent symptoms, vascular flow augmentation may be considered with endovascular procedures, such as angioplasty or stenting [57, 58], or surgical intervention with extracranial-intracranial bypasses.

QMRA pre- and post-endovascular intervention offers useful clinical information that can be followed over time. QMRA of five patients with stenotic basilar or vertebral arteries showed pre- and post-angioplasty flow rates improved by 46 mL/min [59]. QMRA has been used to monitor progress in a patient following treatment of subclavian steal syndrome [60]. In a series of nine patients with intracranial stenosis, QMRA allowed measurement of cerebral hemodynamics following intracranial Wingspan stent placement and found mean volume flow rate in the stenotic artery increased from 81 to 133 mL/min or by 64.2% after stenting [61]. A retrospective study of 18 patients with symptomatic extracranial carotid stenosis demonstrated a 43% improvement of ICA flow post-stenting [62]. Furthermore, QMRA allows the practitioner to detect in-stent restenosis [63, 64] over time.

*2.5.2 Moyamoya Disease*

Moyamoya disease causes progressive occlusion of the supraclinoid internal carotid artery and the middle, anterior, and less frequently posterior cerebral arteries, carrying the risk of stroke. Blood flow may be partially reconstituted by compensatory collateral development or from the posterior circulation. QMRA provides a tool to further assess, characterize, and quantify hemodynamic changes in this disease [65]. In a retrospective study of 66 patients with moyamoya disease, QMRA demonstrated that diseased vessels had lower blood flow, correlating with angiographic staging, and also more robust flow in posterior cerebral and basilar arteries with disease severity [66]. Another study of 35 patients with moyamoya disease found a significant decreased blood flow within the internal carotid artery circulation with a compensatory increase in blood flow in the basilar artery at 2.5 times the normal basilar artery blood flow [67]. Although studies have described alterations in blood flow in moyamoya disease, no studies have yet examined such observed QMRA changes in relationship to stroke risk.

QMRA provides flow measurements before and after surgical interventions, including direct and indirect revascularization procedures, and may be followed over time. This follow-up evaluation becomes particularly relevant in moyamoya disease. In a retrospective study of 62 patients who underwent extracranial-intracranial bypass for all causes including moyamoya, QMRA showed not only patency but a measure of bypass function and found that low or rapidly decreasing flow within the bypass indicated a shrunken or stenotic graft [68]. Postoperative hemodynamic changes in moyamoya patients have been observed by QMRA, with increased ipsilateral hemispheric flow [69] and improvement in the pressure drop along the length of the ICA, as a hemodynamic measure assessing surgical outcome [70]. In a retrospective review of moyamoya patients undergoing combined STA-MCA (extracranial-intracranial artery) bypass and EDAS (encephaloduroarteriosynangiosis), a reciprocal relationship between direct STA bypass flow and indirect EDAS collaterals frequently occurred, with a decline in flow through the bypass over time but with the development of prominent indirect collaterals [71].

## 2.6 Limitations of QMRA

Several challenges remain in the use of QMRA. Although the PC MRA scan time for a given vessel is only 1–2 min, the total acquisition time for TOF MRA, slice selection, and individual vessel PC MRA measurement results in a 45-min study for a full assessment of the major intracranial vessels. The time required for image acquisition with QMRA, particularly with the addition of a vasodilatory challenge, not only restricts the type of vasoactive stimulus used, but it can also lead to artifacts or measurement error in the event of patient motion. Furthermore, diminutive vessels or vessels with stents in place are not well visualized. Slow flow and turbulent flow may result in signal loss from the vessel, which can affect the relative accuracy of measurement.

QMRA with NOVA does not provide the user with the ability to perform post hoc measurements, as the NOVA system acquires flow measurements based on the specified coordinates identified after the TOF MRA. In contrast, 4D PC MRA techniques using a phase-contrast vastly undersampled isotropic projection reconstruction (VIPR) acquisition evaluate the entire cerebral vasculature in a single scan and allow for interrogation of any vessel flows during postprocessing [72, 73]. However, the scan time for a VIPR acquisition is limited to a single VENC for each acquisition, which may potentially lead to inaccurate flow measurement depending on the flows in the vessels of interest. Additionally, the postprocessing for 4D PC MR is not widely available.

Lastly, measurement of blood flow by QMRA is limited to large vessels and is thus only an indirect surrogate measure of tissue-level cerebral blood flow. The functional MR imaging (fMRI) techniques discussed in Subheading 3 focus on assessing tissue-level phenomena that reflect HR. Ultimately, a combination of large vessel and tissue-level measurements can serve as complementary techniques in evaluating patients with steno-occlusive cerebrovascular disease.

# 3 Section 2

## 3.1 Overview

Functional magnetic resonance imaging (fMRI) using blood oxygenation level-dependent (BOLD) contrast uses the hemodynamic response of increased neuronal activity to generate positive MR contrast. The loss of this response can be interpreted as a loss of hemodynamic reserve (HR). Hypercapnia also elicits a vasodilatory response in normal cerebral cortex that results in positive BOLD contrast. Similarly, loss of this response also reflects compromised HR. Although such loss of HR can be used for assessment of CVD, the association with stroke risk has not yet been formally established. This section will discuss the fMRI methods that underlie the application to clinical cases in Subheading 4.

## 3.2 Introduction

Cerebral perfusion is autoregulated within the regional metabolic demands of the brain despite fluctuations in systemic blood pressure changes [74–76]. The pressure range of this autoregulatory process of approximately 60–160 mmHg is not static but modulated by still incompletely understood metabolic and neural mechanisms. As the high rate of cerebral metabolism is obligatorily aerobic and oxygen storage does not occur in brain tissue, a continuous supply of blood transporting oxygen bound to hemoglobin within erythrocytes is mandatory. The release of oxygen in the capillaries is facilitated by the allosteric behavior of hemoglobin and modulated by metabolic factors including tissue pH (i.e., $HCO_3^-$ levels) and the tissue partial pressure of oxygen [77]. Structural factors such as capillary dimensions and density also contribute to efficient oxygen delivery to the tissue [78, 79]. The resting metabolic rate of the brain is high, consuming 20% of the total oxygen supplied to the adult body, despite representing only 2% of body weight, and even higher during the first decade of life [80]. This metabolic rate changes very little across the awake and sleep cycle. Regional increases in local neural network activity related to conscious responses result in increased blood flow into those processing areas. These intricate physiological and biochemical dependencies are based on a reserve capacity to modulate blood flow based on regional cerebral metabolic demands independent of systemic conditions. This phenomenon of the coupling blood flow for supply of oxygen to meet cellular metabolic demands is the physiological basis for BOLD contrast used in functional magnetic resonance imaging (fMRI) [81–86]. This capacity is now defined operationally for the purposes of this chapter as HR.

Loss of HR implies the inability of the brain to modulate blood flow to meet its metabolic demands, thereby indicating increased risk for ischemia when demand surpasses supply. Impairment of HR is a continuous function that may be compromised but not apparent clinically until certain other conditions are met. For example, decreasing hematocrit from a gastrointestinal bleed may exacerbate reduced HR resulting from borderline internal carotid artery stenosis leading to subsequent stroke. Many such clinical scenarios can be proposed in the setting of chronic cerebrovascular disease (CVD), underlying a need to manage stroke risk by objective assessment of HR.

The best characterized parameters for stroke risk assessment are metabolic parameters related to the cerebral metabolic rate of oxygen consumption ($CMRO_2$) as derived from 15-O positron emission tomography (PET) studies [87]. The sensitive parameters are the oxygen extraction fraction (OEF) and the cerebral blood volume (CBV), both of which increase when blood flow is reduced by vascular disease. Unfortunately, this technique, although first proposed more than three decades ago, has never entered routine clinical practice due to the short half-life of 122 s for the 15-O radioisotope, requiring an on-site cyclotron next to the PET scanner in a clinical environment [88–94]. Few laboratories around the world have pursued this technique. Elevated OEF in the setting of increased CBV while $CMRO_2$ and CBF are normal are predictive of increased stroke risk, whereas elevated OEF with normal CBV suggests less risk, probably reflecting at least partially preserved autoregulation. Despite the technical challenges, substantial data have been obtained about cerebral metabolism using $CMRO_2$ derived from oxygen extraction fraction (OEF), cerebral blood flow (CBF), and cerebral blood volume (CBV) measurements [95–97]. The BOLD fMRI method has been shown to be comparable to the PET method of measuring hemodynamic reserve reflected as increases in cerebral blood flow from before to after vasodilation with either intravenous acetazolamide or inhalation of $CO_2$ (hypercapnia) [98].

The goal of management of chronic CVD is to prevent future stroke. Treatment options range from noninvasive approaches of aggressive medical management (e.g., control of hypercholesterolemia, hypertension) with lifestyle changes (e.g., cessation of smoking, increased aerobic exercise, weight loss) to more invasive procedures (e.g., angioplasty, stents, carotid endarterectomy, extracranial-intracranial artery (STA-MCA) bypasses, EDAS (encephaloduroarteriosynangiosis)). Assessment of stroke risk allows an evidence-based selection of the appropriate interventions for individual patients. Because this assessment is an ongoing process requiring re-evaluation of progressive disease, longitudinal studies with reproducible parameters are required that are acceptable for the patient without compromising the value of the data.

This section will describe our clinical experience of using functional MR imaging to evaluate HR longitudinally in patients with established cerebrovascular disease.

## 3.3 Methods

### 3.3.1 Task-Based MR Imaging of Brain Function (Regional BOLD Response)

BOLD contrast fMRI is based on the coupling of the increased regional blood flow to the increased metabolic rate of neurons and surrounding glial cells performing a task [99]. The increased $CMRO_2$ requires increased oxygen delivery to the mitochondria from the surrounding capillaries. Oxygen delivery is a passive diffusion process down an oxygen concentration gradient. As the partial pressure of oxygen at the mitochondria is always low, oxygen diffusion can only be increased by increasing the diffusion gradient by increasing the capillary blood oxygen content. The increased blood flow through the capillary decreases oxygen extraction per volume of blood, thereby increasing the oxygen concentration at the capillary to increase the diffusion gradient. The total oxygen extraction per unit time is increased to meet the metabolic demand. The increased blood flow ensures that the blood oxygenation in the capillaries remains higher than in the baseline state as less oxygen is released per volume of blood. As deoxyhemoglobin is paramagnetic while oxyhemoglobin is diamagnetic, the compartmentation of the deoxyhemoglobin in erythrocytes within the capillaries results in differences in magnetic susceptibility between the erythrocytes and the surrounding tissue [100]. This difference in magnetic susceptibility produces small magnetic field gradients in the applied field of an MR scanner. Diffusion of water across these local magnetic field gradients results in detectable proton MR signal loss on appropriately designed proton MR imaging acquisitions (i.e., T2*-weighted images). As the active task increases blood flow to displace erythrocytes containing deoxyhemoglobin by erythrocytes containing oxyhemoglobin along the length of the capillaries, the local field gradients are removed, and this mechanism for MR signal loss is eliminated. Thus, the increased signal of BOLD contrast is observed. The only way for passive diffusion to deliver more oxygen to the mitochondria, where the partial pressure of oxygen is always low, is to increase the oxygenation of the blood along the capillaries and enhance its release from hemoglobin. This is achieved by increasing blood flow beyond that required to increase the oxygen concentration gradient above that of the baseline state with concomitant decrease in pH to displace the hemoglobin-oxygen binding curve to favor the release of oxygen (Bohr effect). BOLD contrast is a regional increase in proton signal in response to increases in neuronal activity and a direct measure of the integrity of neural-vascular coupling and HR. The process is temporally relatively slow following several seconds after neuronal activity has occurred and similarly slow to recover after the task is completed.

Assessment of BOLD contrast simultaneously in each vascular territory provides a simple means to determine if HR is sufficient to support the BOLD response in those regions of cortical activation. The paradigm that we have used in our clinical service for the last 10 years uses a block design, audiovisual bilateral hand grasping task that lasts 4.5 min with 30 s of central fixation alternating with 30 s of hand grasping for 4.5 cycles, starting with fixation and ending with fixation (Fig. 5a) [101]. The bilateral hand grasping condition moves the fingers and thumbs paced at one movement per second by audio commands (OPEN, CLOSE) delivered via pneumatic earphones as well as visual commands displayed centrally on a screen viewed via an angled mirror on the radiofrequency head coil. The robustness of the visual stimulus is increased by displaying alternating flashing (10 Hz) black and white checkerboards on either side of the centrally displayed visual commands (OPEN, CLOSE). The patient must be trained to not make a tight fist, only to move the fingers and thumb to touch the palm of the hand, thereby avoiding excessive effort that causes head motion.

The fMRI acquisition protocol for the scanner uses gradient-recalled echo, echo planar imaging (TR/TE = 2500/25 ms, FOV = 240 × 240 mm$^2$, matrix size = 64 × 64, slice thickness/gap = 3/0 mm, number of slices = 35) that is automatically synchronized with the presentation of the paradigm using the scanner RF trigger (MRIx Technologies, Bannockburn, IL, USA). This hardware also monitors the eye position of the patient by infrared video to provide real-time feedback to the technologist to ensure head position remains stationary. If the patient moves, the technologist instructs the patient to return to the standard head position established during training.

The activation patterns from this paradigm have been compared to MR perfusion before and after acetazolamide-induced vasodilation in a retrospective patient series [101]. Poor concordance was posited to be due to the leptomeningeal collateral flow producing prolonged mean transit times that further increased after vasodilation due to longer path lengths and decreased pressure gradients between vascular territories but still with adequate flow to maintain sufficient HR to support BOLD contrast [101]. We discontinued the use of perfusion studies with this realization and replaced MR perfusion with the hypercapnia paradigm discussed in detail below.

The magnitude of the BOLD signal change increases with magnetic field strength increasing from 1% to 2% at 1.5 T to 2–4% at 3.0 T [102]. Even at 3.0 T, such small signal changes remain below visually detectable thresholds for even well-trained neuroradiologists. Signal averaging and statistical detection methods are required to isolate signal differences between the active and baseline conditions. This detection assumes minimal head motion between images acquired between the two conditions. The patient must be cooperative and trained to perform this paradigm prior to entry into the scanner on a laptop computer. Two trials of the

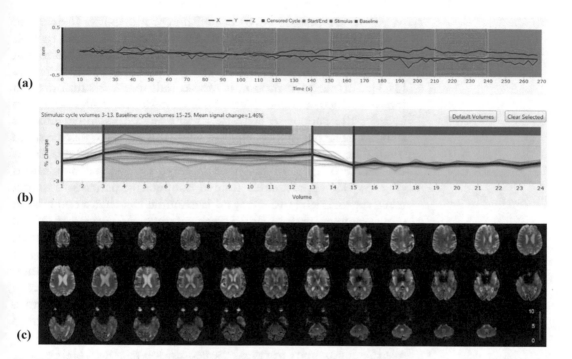

**Fig. 5** (a) Schematic of the audiovisual-paced bilateral hand grasping paradigm with a block design of 4.5 cycles (270 s, 108 volumes) of the baseline condition (blue, 30 s) and active condition (green, 30 s) with a start condition (red, 30 s). The translational motions of the center of mass along the three orthogonal axes (X, blue; Y, green; Z, red) are superimposed over the block-designed paradigm. Manual processing, if needed, uses the motion plot to censor any cycles in which head motion is excessive. This is empirically motion of more than about 1/3 of the voxel dimension (3.125 mm × 3.125 mm in-plane and 3.0 mm through-plane). If head motion is gated to the paradigm, then the data must be discarded and the paradigm repeated, after again emphasizing the importance of head stability to the patient. (b) The automated analysis performs a voxel-wise statistical comparison of the average signal intensities between active and baseline conditions averaged over all four cycles. The statistical comparison can be performed with a Student $t$-test without or with a Bonferroni correction or with a false discovery rate. The start condition is not used in the analysis as this time establishes cognitive stability of the patient after the noisy scanning starts and spin equilibrium for the MR imaging. The mean BOLD signal change between conditions is calculated automatically for selected voxels (e.g., 1.46% shown for voxels selected from motor cortex). Manual processing also permits selection of the number of images in each condition to be used in the statistical comparison. The separation of the images between the two conditions can be changed if the BOLD response is delayed. The automated comparison shown omits two volumes (5 s delay between active and baseline conditions) to avoid the transition in BOLD contrast between the two conditions. (c) A mosaic display during manual processing allows the full activation pattern across the brain to be appreciated. This clinical paradigm has internal control activation with activation in the cerebellum as well as in the cerebrum for the motor activation. If cerebellar activation supported by the posterior circulation is present but primary motor activation supported by the anterior circulation is missing, the patient is performing the paradigm, but HR must be insufficient to support BOLD contrast. In this case of a normal volunteer, activation is present throughout the primary sensorimotor cortex along the pre- and post-central gyrus bilaterally (superior middle cerebral arteries), supplementary cortex along the medial frontal lobes (anterior cerebral arteries), auditory cortex along the superior temporal gyrus bilaterally (inferior middle cerebral arteries) and along the calcarine fissure in the occipital lobes (posterior cerebral arteries). The color scale represents the $t$-threshold for the Student $t$-test detection of voxel-wise signal differences between the two conditions of the block design audiovisual motor paradigm

paradigm are performed to establish reproducibility. If excessive head motion is detected, the paradigm is repeated. Head motion is minimized using a visual feedback system (MRIx Technologies, Bannockburn, IL, USA) which the patient is also trained to use [103]. Further head motion is monitored by eye position with an infrared video camera. This video allows the technologist to instruct the patient to stabilize their head in real time into the standard head orientation established by the visual feedback system. Typical head motion is less than ±0.5 mm translational motion of the center of mass of the head in three orthogonal dimensions (Fig. 5a) for appropriately trained cooperative patients. Training takes approximately 15 min and is essential to achieve compliance with resultant high-quality activation maps. If necessary, visual acuity correction with MR-compatible lenses is used if the patient is not using contact lenses. Acquisition of high-quality data not compromised by head motion is emphasized. Postprocessing re-registration of images across the time series does not improve results and is equivalent to blurring the voxel-wise data with a spatial filter. Real-time monitoring of compliance with stable head position allows assessment of data quality and allows the technologist the opportunity to repeat the acquisition to obtain high-quality data immediately.

The activation maps are processed automatically for each of the two trials of the paradigm after the time series of images are transferred offline from the scanner to a workstation with a commercial software package (MRIx Technologies, Bannockburn, IL, USA). For each trial, the two conditions of the paradigm are averaged voxel-wise across the four cycles of the trial. Then, selected portions of the two averaged conditions are compared statistically by Student $t$-test for BOLD contrast (Fig. 5b). The BOLD signal change between the two conditions of the paradigm has a delayed temporal response of several seconds that may also be modified by pathology and should be considered during analysis and interpretation. In normal individuals, the BOLD delay is usually about 5 s or two volumes of data when using a repeat time TR = 2.5 s. The transition period (5 s, two volumes) between the two conditions is omitted, and the selected portions can be adjusted for the delayed BOLD response. The first half cycle of the paradigm is not included in the analysis, as this establishes the cognitive stability of the patient as well as the proton longitudinal spin equilibrium. Any cycles compromised by head motion are removed to improve map quality. A mosaic of all the axial partitions of the activation map (Fig. 5c) is reviewed during manual processing to allow the pattern of activation to be appreciated. Any head motion is easily detected by the false activation around the high contrast borders of air-bone-tissue interfaces. This paradigm is very robust with at least 50% redundancy so that only two cycles of each trial are required to still obtain a useable map. Statistical thresholds to distinguish BOLD contrast from noise are chosen based on the scanner performance (1.5 or 3.0 T) and any reduction in number of cycles censored because of head motion. Although a positive tail Student $t$-test is usually used, Bonferroni correction and false discovery rate thresholding are also available and yield similar results.

The normal activation pattern for a single trial is displayed for selected partitions in Fig. 6 with labels of each activation node. The normal pattern shows activation in the primary sensorimotor cortex on both sides of the central sulcus bilaterally, along the standard neurosurgical landmark (i.e., the knob or inverted omega sign). Right-handed patients show greater activation for the non-dominant hand, and this asymmetry should not be taken to reflect an abnormality. There is often activation along the precentral sulcus bilaterally due to eye movement reflecting the frontal eye fields. These regions are supplied by the superior middle cerebral arteries. The corresponding motor areas of the cerebellum should be activated when these areas activate and constitute the necessary

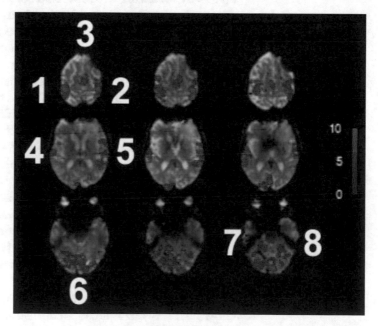

**Fig. 6** Selected partitions from Fig. 5 labeling the areas of activation present in a normal activation pattern when HR is intact. 1 and 2 = right and left sensorimotor activation for hand grasping and eye movement anterior and posterior to the central sulci, respectively; 3 = bilateral supplementary motor activation along the medial frontal lobes; 4 and 5 = right and left auditory activation along the superior temporal gyri, respectively; 6 = bilateral primary and association visual activation along the calcarine fissures bilaterally; and 7 and 8 = right and left cerebellar motor activation in the cerebellar hemispheres, respectively. This numbering scheme is used for all subsequent figures. The color scale at the right is the positive $t$-threshold of the Student $t$-test ($t > 3$) used to discriminate BOLD activation from noise

control observation to reliably interpret the loss of activation as due to HR compromise in the superior branches of the middle cerebral artery. The supplementary motor areas along the medial frontal lobes are supplied by the anterior cerebral arteries and can be asymmetrically located in both the anterior-posterior and superior-inferior dimensions. The auditory cortex along the superior temporal gyrus bilaterally is supplied by the inferior branches of the middle cerebral arteries. The primary and associative visual cortices along the calcarine fissures in the occipital lobes are supplied by the posterior cerebral arteries. This region of the brain is closest to the head support in the supine patient and furthest from the paranasal sinuses and petrous bones so that this activation is robust and usually the most active across the pattern of activation and can be used to set statistical thresholds appropriate to the data quality when other regions may not be present.

When the data are processed automatically, the data from each trial must be inspected for artifacts and appropriate thresholding (Fig. 7) by reviewing the activation superimposed over the original echo planar images from which it is derived (Fig. 7a), over the T2-weighted fast gradient-recalled echo images acquired immediately prior to the trial when head motion should be minimal (Fig. 7b), and as a mosaic to appreciate the global pattern of the activation (Fig. 7c). The activation map superimposed over the high-resolution 3D T1-weighted images should also be reviewed in three planes (Fig. 7d, left, axial; middle, coronal; right, sagittal) to ensure that the activation is localized accurately in three planes. This is important to avoid mistaking noise for activation (false positive) when areas of activation are missing due to compromised HR when the thresholding is lowered.

The MRA and $CO_2$ reactivity maps (Fig. 8a–c) should be reviewed and cross-referenced to the task-based BOLD activation maps from the two trials. This is simplified by superimposing both activation patterns over 3D high-resolution anatomic T1-weighted images in three orthogonal planes (Fig. 8d–f). Each activation map is displayed in its own selectable color (usually green for the first trial and magenta for the second trial) with the overlap colored in yellow. The overlap areas demonstrate the robustness of the regional BOLD activation and the integrity of the HR.

### 3.3.2 MR Imaging of Vascular Reactivity to Hypercapnia (Global BOLD Response)

Whereas the regional BOLD activation maps examine the HR in specific cortical regions of motor, auditory, and visual processing, the hypercapnia challenge is designed to measure the vascular response to elevation of $CO_2$ across all brain regions. Although several hypercapnia inhalation techniques for assessment of vascular reactivity have been reported and commercial products can perform controlled gas delivery [104–110], we have elected to use an inexpensive strategy that involves only disposable components available from our respiratory therapy service and expired carbon dioxide.

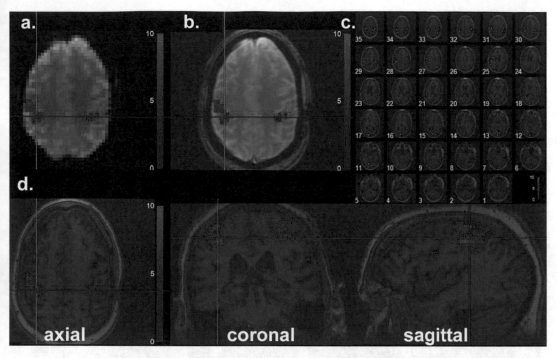

**Fig. 7** Systematic review of each activation map is performed by reviewing (**a**) activation map superimposed over the original axial image data (3 T, 8-channel head coil, T2*-weighted gradient-recalled echo, echo planar images: TR/TE = 2500/30 ms; FOV = 240 × 240 mm², 64 × 64 mm²; 3/0 mm slice thickness/gap, 34 slices) to isolate areas of reduced signal intensity from bulk magnetic susceptibility effects at bone-air interfaces and arising from surgical sites. The same color scale of the *t*-threshold of the Student *t*-test used to detect BOLD contrast is used in (**a**)–(**d**) panels. (**b**) Activation superimposed over axial higher-resolution T2*-weighted gradient-recalled echo images (TR/TE = 500/20 ms, FOV = 240 × 240 mm², matrix size = 256 × 160, slice thickness/gap = 5/1 mm, 35 slices) co-registered by isocenter and slice thickness to the fMRI echo planar images and acquired immediately prior to the functional echo planar imaging acquisition to minimize any head motion that may occur over extended times between the functional acquisition and high-resolution images. (**c**) Mosaic of all axial activation partitions in a single matrix arrangement with a background of high-resolution T1-weighted gradient-recalled echo images (TR/TE = 9.328/3.828 ms, FOV = 240 × 240 mm², matrix size = 320 × 256, slice thickness/gap = 1.5/0 mm, 100 slices) used to provide a global view of the pattern of activation, readily revealing any variations from the normal pattern. Images are labeled 1 through 35. (**d**) Cross-referenced display of activation on three cross-referenced orthogonal planes (axial plane = green lines, coronal plane = red lines, sagittal plane = yellow lines) through high-resolution T1-weighted gradient-recalled images (3 T, 8-channel head volume, TR/TE = 9.328/3.828 ms, FOV = 240 × 240 mm², matrix size = 320 × 256, thickness/gap = 1.5/0 mm, 100 images). The displayed cross-referenced partitions are shown as intersecting the right sensorimotor cortex activation

A closely fitting small face mask, that covers the face of the patient positioned within the 8-channel head coil, is connected to the supply of oxygen gas available in all MR magnet rooms. The mask is also connected to a flexible expandable respiratory tube that can extend to a volume larger than the end-tidal volume and is open to the atmosphere. If the mask is too tight for the patient's head within the head coil, the mask can be trimmed with scissors. The

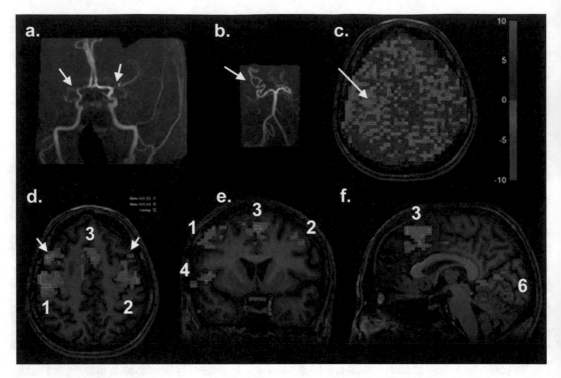

**Fig. 8** Example of a patient with a moyamoya variant being evaluated for HR. (**a**) The MR angiogram of the anterior intracranial circulation shows bilateral occlusions of the M1 segments of the middle cerebral arteries (arrows) with intact anterior cerebral arteries and supraclinoid portions of the internal carotid arteries. (**b**) The posterior circulation shows a normal basilar artery with an expanded vascular territory for the right posterior cerebral artery (arrow), suggesting compensating leptomeningeal flow. (**c**) A single representative axial partition from the vasoreactivity map showing a largely intact hypercapnia response slightly reduced in the right frontal lobe (arrow). (**d–f**) The regional BOLD activation map for the audiovisual bilateral hand motor paradigm performed twice (activation from the first trial in green and second trial in magenta) has the overlap of activation areas in yellow to show reproducibility. The sensorimotor (1, 2) and supplementary motor (3) areas show reproducible regional BOLD activation in the axial (**d**), coronal (**e**), and sagittal (**f**) planes. The visual cortex (6) is not well displayed on the midline sagittal plane (**f**). There is also activation of the frontal eye fields in the precentral sulci (**d**, arrows). The activation maps are cross-referenced on the supplementary motor area (3) in three orthogonal planes (**d** = axial, **e** = coronal, **f** = sagittal). The background images are high-resolution T1-weighted gradient-echo images co-registered at identical isocenter with the functional images. These data are transferred to PACS for display, review, and interpretation on the clinical service

$CO_2$ level is monitored in the face mask (Invivo MRI Patient Monitor, a Philips Healthcare Company, Gainesville, FL). The hypercapnia paradigm (Fig. 9) consists of three cycles of alternating between the condition of oxygen flowing (3 L/min, 1 min) into the mask purging the expired $CO_2$ out of the mask and tube, that are open to the atmosphere, and the condition of hypercapnia when no oxygen is flowing into the mask, allowing expired $CO_2$ to accumulate in the mask and tube (2 min). The extended hypercapnic condition of 2 min is required because of the delayed vascular reactivity response that would otherwise not be detected in patients

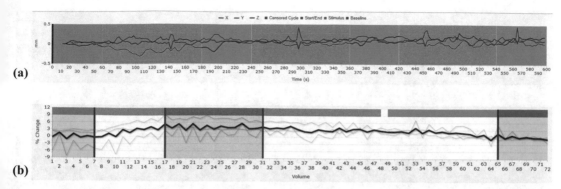

**Fig. 9** (a) Schematic of the three-cycle $CO_2$ challenge paradigm showing a start condition (red) of flowing oxygen with low $CO_2$ followed by three cycles of rebreathing expired $CO_2$ without flowing oxygen (green, 2 min) followed by flowing oxygen purging out expired $CO_2$ (blue, 1 min). The head motion along three translational axes ($x$ axis = blue, $y$ axis = green, $z$ axis = red) is displayed across the paradigm and is less than $\pm 0.5$ mm as shown on the vertical axis. The duration of the three cycles and start condition is 10 min as shown on the horizontal axis. (b) The voxel-wise analysis comparing the two conditions (green vs. blue) is performed automatically averaging the three cycles. The selection regions can be varied manually to fit the temporal response of the BOLD MR signal change which increases in hypercapnia. The averaged response can be verified manually for each cycle as with the task-based fMRI analysis as one cycle is robust enough to show a statistically robust response. $CO_2$ monitoring during the image acquisition shows large changes in the $CO_2$ levels (partial pressure of $CO_2$ varies systematically from around 17 up to about 45 mmHg across each cycle). Arterial blood oxygenation by peripheral arterial oximetry remains at full saturation. The patient is not aware of any significant changes

with severe disease. The oxygen gas ensures that the patient remains at 100% arterial saturation as monitored by peripheral pulse oximetry on the same monitoring device. The mask must fit the face perfectly and can be achieved using surgical tape and gauze. During the hypercapnia condition, the patient breaths their own expired gases that remain in the mask and tubing. As the tubing is open to the atmosphere, there is no pressure change or resistance to breathing. However, the $CO_2$ level changes from low values (partial pressures of $CO_2$ ~15 mmHg when $O_2$ is flowing) to elevated values (partial pressures of $CO_2$ ~45 mmHg when oxygen is not flowing) to elicit a robust vasodilatory response in the normal brain. The hypercapnia paradigm is 10 min in duration beginning with 1 min of flowing oxygen followed by three cycles of alternating hypercapnia and normocapnia that is equivalent to three repeated $CO_2$ challenges. Imaging covers the entire brain to allow detection of BOLD contrast across the entire cerebral cortex and cerebellum. Vascular response maps are automatically processed using the same software as for regional BOLD activation maps (as previously shown in Fig. 7) and must be scrutinized for artifacts and motion in the same way (Fig. 10). However, the maps are presented showing both positive (red-yellow color scale) and negative (green-blue color scale) statistical parameters to display the normal

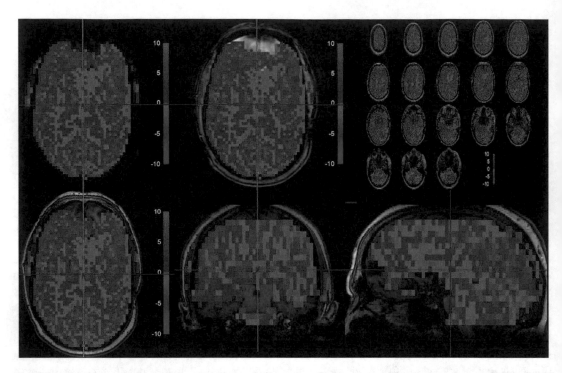

**Fig. 10** A normal vascular reactivity map shown in the same display as used for the task-based fMRI in Fig. 7 showing the robust symmetric hypercapnia response throughout the cortex of both cerebral hemispheres with reduced but still positive response in the white matter, obtained using the paradigm shown in Fig. 9a. Unlike the task-based analysis, both sides of the *t*-test are displayed to detect intact reactivity (red to yellow scale) and loss of reactivity (green to blue scale). For this normal patient, there is no loss of reactivity. These data are transferred to PACS for display, review, and interpretation. Each cycle can be manually processed to show the reproducibility of the reactivity of individual cycle's responses albeit with less statistical significance

positive vasoreactive response (+BOLD response with increased $T2^*$-weighted MR signal during elevated $CO_2$) and the pathological response (-BOLD response with decreased $T2^*$-weighted MR signal during elevated $CO_2$). The abnormal response (blue brain) is interpreted as implying further desaturation of capillary blood in compromised tissue distal to a stenosis. Oxygen extraction is further increased as blood flow to these regions is reduced when vascular resistance in neighboring tissue without proximal stenosis is decreased. The resultant reduction in pressure gradient between vascular territories redistributes blood flow, reducing leptomeningeal collateral flow. This abnormal response will be illustrated under the clinical cases. These vascular reactivity maps can also be displayed over 3D high-resolution anatomic T1-weighted images of the brain and displayed in three orthogonal planes as with regional BOLD task-based maps.

The acquisition protocol on the MR scanner is the same as that used for the task-based fMRI protocol, except longer in duration. Manual processing of the image data allows the three cycles of the paradigm to be compared to establish reproducibility as each cycle contains sufficient data for robust individual analysis. If head motion compromises one cycle, these data can be eliminated. The patient is required to remain stationary for 10 min but does not experience any effect of the changes in composition of the inhaled gases. The visual feedback system, as used for the task-based fMRI, is maintained for the entire 10 min which ensures stable head position in cooperative well-trained patients.

In practice, this $CO_2$ challenge paradigm is used before the task-based fMRI paradigm. The mask is removed after the $CO_2$ challenge to avoid prolonged discomfort from the mask and tape over the face. This requires two separate 3D high-resolution T1-weighted image datasets to be acquired as the head inevitably moves when the patient is moved out of the magnet to remove the mask.

### 3.4 Discussion

In the past, dynamic susceptibility contrast (DSC) MR perfusion studies before and after vasodilation with acetazolamide were performed in parallel with task-based BOLD fMRI studies. That clinical experience showed that the perfusion and BOLD fMRI results were not concordant [101]. The prolonged mean transit times (MTT) distal to severe proximal stenoses were attributed to leptomeningeal collateral flow that could still be adequate to maintain regional BOLD contrast. Rather than implying compromise of tissue perfusion, prolonged MTT could also imply adequate, albeit delayed, collateral flow. The negative BOLD contrast observed during the hypercapnia challenge in cases of severe proximal stenosis of the cerebral arteries can be attributed to increased oxygen extraction from the prolonged blood flow through the compromised tissue. We replaced the perfusion studies with the hypercapnia paradigm to ensure that all regions of the cortex are interrogated for intact vascular reactivity. Whereas the task-based fMRI approach relies on neural-vascular coupling in a neuronal stress test, the hypercapnia challenge is the equivalent to the pharmaceutical challenge with the carbonic anhydrase inhibitor, acetazolamide, without significant change in neuronal activity. As such, we regard these two methods to be independent and complementary measures of HR integrity.

To illustrate the methods discussed in Subheadings 2 and 3, a set of case studies are presented below.

## 4  Section 3

### 4.1 Case Studies

The following cases serve as examples of the roles that QMRA and fMRI can play in the management of patients with complicated CVD. All imaging has been performed on our clinical MR scanners operated by clinical technologists as an integral part of the clinical MRI service. All images have been captured from PACS and clinical workstations.

This 21-year-old left-handed female student presented with tingling in hands and legs for over 6 months with episodes of slurred speech lasting about 1 min. Initially, she underwent MR imaging for clinical concern for multiple sclerosis or Lyme disease at an outside institution. A subsequent cerebral angiogram confirmed the diagnosis of moyamoya disease. Continued progression of symptoms resulted in the referral for possible surgical intervention.

The MR imaging evaluation of cerebrovascular pathophysiology included anatomic MRI and MRA examinations followed by evaluation of HR using functional MRI with task-based and hypercapnia paradigms and quantitative flow measurements of the extracranial and intracranial circulation before and after provocative vasodilation. Preoperative MRI evaluation showed small chronic ischemic changes within the white matter of the brain parenchyma bilaterally but worse in the left frontal lobe with lacune formation but no evidence of acute or early subacute ischemic changes (Fig. 11a–c). These chronic changes may have been due to emboli

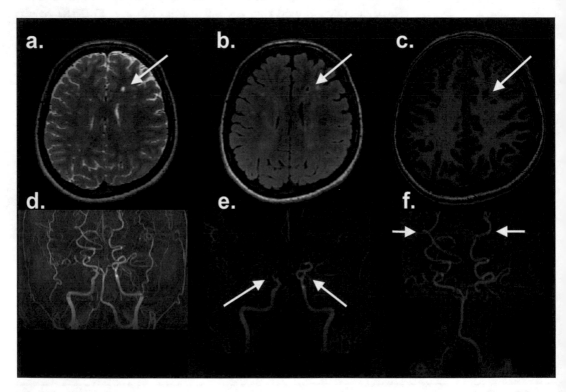

**Fig. 11** (a) Axial T2-weighted fast spin echo, (b) T2-weighted FLAIR, and (c) T1-weighted spin echo images through the brain showing chronic ischemic changes and lacunae formation (arrows). (d) Maximum intensity projection reconstruction of the 3D time-of-flight MR angiography of the entire intracranial circulation. (e) Intracranial anterior circulation angiogram showing obstruction of the supraclinoid portions of the internal carotid arteries bilaterally without normal flow enhancement in the proximal anterior or middle cerebral arteries bilaterally (arrows). (f) Intracranial posterior circulation showing expanded posterior cerebral arteries bilaterally (arrows). Such findings are classical moyamoya disease

from the abnormal vessels at the circle of Willis or reflect low flow conditions insufficient to meet the metabolic demands of the distal tissue. The MRA examination showed the moyamoya configuration of the anterior intracranial circulation with occlusion of the supraclinoid portions of the internal carotid arteries extending into the A1 and M1 segments bilaterally with multiple small vessels around the circle of Willis (Fig. 11d, e). The posterior circulation showed expanded vascular territories of the posterior cerebral arteries bilaterally indicating leptomeningeal collateral circulation (Fig. 11f). However, this information remained ambiguous as to whether flow into the anterior circulation was adequate.

HR, assessed with the fMRI protocol (Fig. 12), showed bilateral loss of vascular reactivity in the anterior circulation (Fig. 12a, b) with chronic ischemic changes (Fig. 12c and also shown in Fig. 11) with no regional BOLD response in the sensorimotor and supplementary motor areas as well as auditory cortex (Fig. 12d, e), despite adequate task participation by the patient, as indicated by the normal regional BOLD activation of the visual and cerebellar motor response(Fig. 12e, f). These results can be interpreted as indicating loss of sufficient HR to support regional BOLD contrast in the anterior and middle cerebral arteries bilaterally, despite the leptomeningeal collateral circulation from the posterior cerebral arteries reflected by the MR angiography.

Further HR assessment used quantitative flow measurements before and after vasodilation with intravenous acetazolamide (1 g for adults without contraindications). These results are shown in Fig. 13.

Based on the anatomic changes, the HR results from fMRI and QMRA, and the clinical assessment, flow augmentation by a left superficial temporal artery to distal middle cerebral artery bypass with encephaloduroarteriosynangiosis (EDAS) was elected. The left side was chosen based on the greater anatomic changes and the greater dependence on leptomeningeal flow from the left posterior cerebral artery. The left internal carotid artery supplemented the flow from the basilar artery through the left posterior communicating artery. Follow-up evaluation after the bypass showed that the bypass remained intact but with greater worsening of HR on the right side. Figure 14 shows the MR imaging assessment at about 7 weeks following surgery. The quantitative flow measurements were also repeated before and after vasodilation with acetazolamide (Fig. 15).

The time course of signal changes during the task-based fMRI study in the compromised vascular territories of this patient is significantly displaced temporally compared to normal regions of the brain, as shown in Fig. 16. The BOLD response is usually delayed after switching conditions by about 5 s or two acquisition volumes for a TR = 2.5 s, as seen in visual cortex in Fig. 16a.

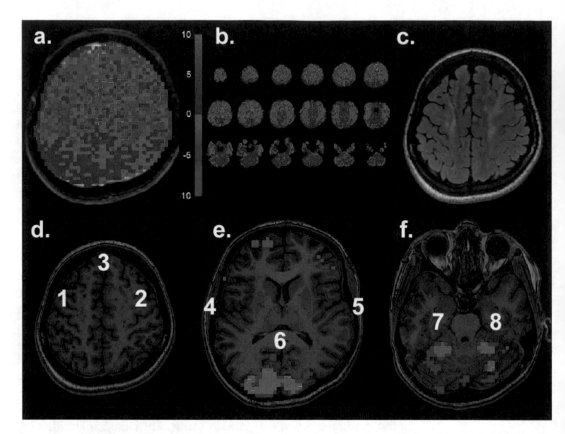

**Fig. 12** Preoperative assessment of HR by fMRI with the hypercapnia challenge shows (**a**) axial partition through the motor cortex and (**b**) mosaic of all axial partitions through the vascular reactivity map showing inadequate reactivity (blue) to increase blood flow in response to hypercapnia throughout the anterior circulation, whereas the posterior circulation retains the capacity for vasoreactivity (red). The color scale (−10 to +10) reflects *t*-values for comparison of signal intensities in the hypercapnia and normocapnic conditions of the paradigm. The positive (yellow) scale indicates an increase in BOLD contrast from the baseline condition. The negative (blue) scale indicates an abnormal decrease in BOLD contrast from the baseline condition. (**c**) Axial T2-weighted FLAIR image through the same location as (**a**) to show the corresponding anatomy. (**d**) No regional BOLD activation is detected in the sensorimotor (1, 2) and supplementary motor areas (3) or (**e**) auditory cortex (4, 5) despite normal responses in visual cortex (6) and (**f**) bilateral cerebellar motor areas (7, 8). These two independent approaches indicate that HR remains intact only in the posterior circulation but is inadequate to support regional BOLD responses and vascular reactivity in the anterior circulation. Numbers are defined as in Fig. 6

However, the response was further delayed in compromised vascular territories such as motor cortex, as shown in Fig. 16b, even though the bypass had restored sufficient HR to support BOLD contrast. This is not unexpected given that the vascular reactivity (Fig. 14b) has not returned to a normal response.

Concern for persistent HR compromise (Fig. 14b) indicated the need for further flow augmentation, and a second similar bypass and EDAS were performed on the right side. Follow-up MR imaging performed at about 8 weeks after surgery is shown in Fig. 17.

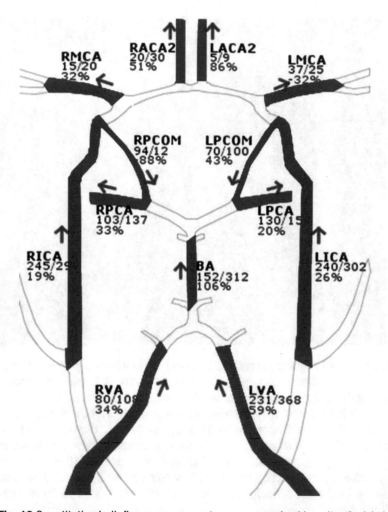

**Fig. 13** Quantitative bulk flow measurements are summarized in units of mL/min through each of the intracranial vessels (as shown in Fig. 1). The flows through the anterior and middle cerebral arteries are considerably lower than expected for the patient's age with no significant increase and some vessels showing a decrease following vasodilation indicating compromised HR. In contrast, the posterior cerebral arteries, basilar artery, vertebral arteries, posterior communicating arteries, and internal carotid arteries all increase flows following vasodilation, indicating intact HR. The baseline flows in the posterior circulation are higher than expected for age reflecting compensating leptomeningeal collateral flow to the anterior circulation

Improved vascular reactivity and new regional BOLD activation indicated that HR had been improved.

Although the hypercapnia response had returned toward normal, timing delays persisted between the original compromised and normal vascular territories compared to normal brain previously illustrated in Fig. 9. Such temporal delays, as shown in Fig. 18a, b for this patient, were present in both regions of positive (yellow)

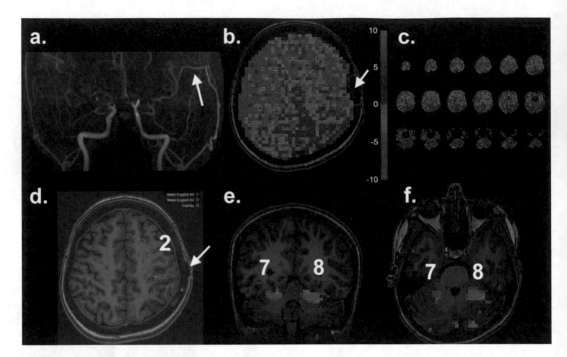

**Fig. 14** (a) MR angiography shows the patent left-sided bypass (arrow) that was not present on the angiogram in Fig. 11d. (b) Magnified single partition through the sensorimotor and supplementary motor cortex selected from (c) the entire mosaic of axial partitions from the vascular reactivity map from the hypercapnia challenge fMRI showing worsening (blue compared to Fig. 12a, b) over the right anterior circulation but also in the left anterior circulation despite the left-sided bypass (arrow). The regional BOLD activation map (d), however, demonstrates restored activation in the left sensorimotor cortex (2) adjacent to the left-sided bypass (arrow) not present on the preoperative study (Fig. 12d). (e) Coronal and (f) axial images demonstrate cerebellar motor activation (7, 8) indicating that the patient was performing the task adequately. Auditory cortex was not detected, but visual activation was present and unchanged from the preoperative study (not shown but similar to Fig. 12e)

and negative (blue) vascular reactivity. These delayed responses required that the hypercapnia condition of the paradigm be at least 2 min in duration. These delays suggest that the redistribution of blood flow induced by the hypercapnic changes in vascular resistance is delayed due to the longer time taken for blood to transverse the vasodilated regions with normal vascular reactivity. The longer time to transverse the leptomeningeal vessels into the regions with compromised vascular reactivity reflects the reduced pressure gradients along these collateral vessels between the normal and compromised vascular territories. Such a model has been proposed elsewhere to explain why dynamic susceptibility contrast perfusion studies are not concordant with BOLD fMRI studies [101]. The increased oxygen extraction in the compromised regions ensures that the magnitude of the abnormal BOLD contrast is as easily detected as normal BOLD contrast.

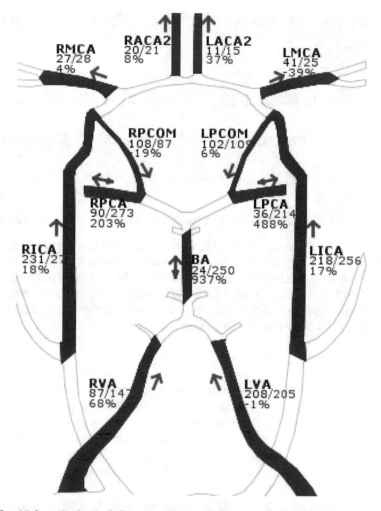

**Fig. 15** Quantitative bulk flow measurements following the left-sided bypass are similar to the preoperative study (Fig. 13) with a bypass flow of about 50 mL/min with no increase following vasodilation (not shown on the diagram). The flows through the proximal anterior and middle cerebral arteries remain low without significant change with vasodilation as would be expected. The posterior cerebral arteries and basilar artery show lower baseline flows but greater flows following vasodilation compared to the preoperative study. Concern remained that HR was still not adequate in the right cerebral hemisphere

This case illustrates that the loss of HR can be assessed using fMRI strategies for the initial evaluation of a patient with severe cerebrovascular disease and used to monitor disease progression and response to interventions. Quantitative measurements of bulk flow in the large vessels prior to and after vasodilation can establish the presence of leptomeningeal collateral flow. The two fMRI approaches are independent but complementary methods. The regional BOLD contrast method relies on the metabolic-vascular coupling of neuronal activity, i.e., the equivalent of a neuronal stress

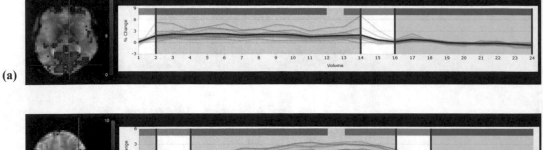

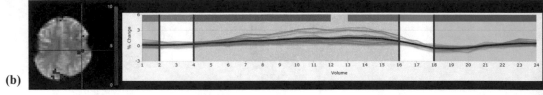

**Fig. 16** (a) Time course of BOLD contrast in multiple voxels from the visual cortex which shows a normal delay of 5 s between the active condition (green) and the baseline condition (blue) for the task-based audiovisual paradigm. Mean BOLD contrast is as high as 6%. (b) Time course of BOLD contrast in multiple voxels from the newly restored left motor cortex following left superficial temporal artery-middle cerebral artery bypass shows a latency of 10 s (decreasing at volume 16 rather than 14) compared to visual cortex (a). Mean BOLD contrast is about 2.2% in the recovered activation

test. The hypercapnia vascular reactivity method is the equivalent of a pharmaceutical challenge detecting regional variations in a global response. Both methods depend on the integrity of the HR to manifest the normal patterns of BOLD contrast.

Although this case of moyamoya disease shows concordance for the fMRI and QMRA methods with loss of HR, this is not always the case, as illustrated by the next cases.

*4.1.2   Case B: Moyamoya Disease (Flow-Compensated Intracranial Circulation)*

This 37-year old right-handed woman presented with left-sided weakness suggestive of stroke after occasional episodes of bilateral lower extremity weakness. Evaluation with MRA (Fig. 19) found a focal right M1 segmental stenosis, high-grade stenosis of the proximal left middle cerebral artery, and focal stenosis of the origin of the left A1 segment. There was also indication of a left extracranial stenosis of the left internal carotid artery.

The results of the fMRI evaluation of HR in Fig. 20 showed no significant sequelae of ischemic disease despite the symptoms at presentation and the MR angiogram results showing a moyamoya configuration of the intracranial circulation (Fig. 19). The task-based fMRI shows BOLD activation in all vascular territories. The vascular reactivity was reduced in the white matter of both cerebral hemispheres, but the cortex was not severely compromised. The QMRA results (Fig. 21) confirm the increased flow through the left anterior and posterior cerebral arteries indicating increased leptomeningeal collateral flow. This patient did not proceed to bypass surgery but is undergoing aggressive medical management.

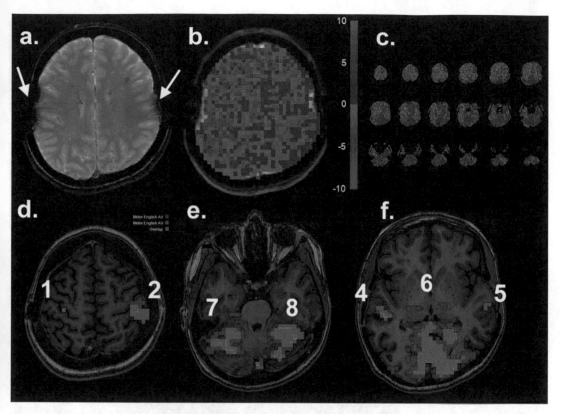

**Fig. 17** (**a**) Axial T2*-weighted gradient-recalled imaging shows the bilateral bypasses by the susceptibility artifacts arising from the craniotomies (white arrows). (**b**) Magnified single partition through the sensorimotor cortex from (**c**) the mosaic of axial partitions of the vascular reactivity map shows improved vascular reactivity but still asymmetric with less severely reduced reactivity (more blue) on the right side compared to the pattern following the first bypass procedure (Fig. 16b, c) and preoperatively (Fig. 12a, b). (**d**) One partition of the axial regional BOLD activation map (superimposed over T1-weighted gradient-echo image) shows new asymmetric BOLD activation in the right (1) and left (2) sensorimotor cortex, while (**e**) bilateral activation in the cerebellar hemispheres (7 right, 8 left) verifies that the patient was adequately performing the task bilaterally. (**f**) Regional activation is shown in the auditory cortex along the superior temporal gyri (4 right, 5 left) and visual cortex (6). Activation in the bilateral auditory and right sensorimotor areas had not been detected previously for this patient. The supplementary motor areas are still not visualized

The MR imaging protocol for re-evaluation of HR is not only useful in such complicated entities as moyamoya disease but can also be useful to manage artherosclerotic and other steno-occlusive diseases, as shown in the next case.

*4.1.3 Case C: Left Internal Carotid Artery Stenosis (Flow-Compensated Intracranial Circulation)*

This 21-year-old right-handed woman presented with right-sided weakness and numbness which spontaneously resolved. Left internal carotid artery dissection was suspected, but cerebral angiography showed left internal carotid artery stenosis but not typical of dissection. There were two subsequent episodes of dizziness with visual blurring, one severe enough to cause loss of consciousness. The MR imaging evaluation (Fig. 22) showed no tissue injury

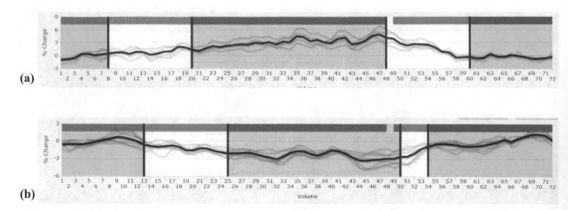

**Fig. 18** BOLD contrast (% change) voxel-wise time course across the hypercapnia condition (green) and normocapnic condition (blue) averaged for the three challenges in voxels within (**a**) regions of positive vascular reactivity (yellow brain, increased signal change at the end of the hypercapnia challenge) and (**b**) regions of negative vascular reactivity (blue brain, decreased signal change at the end of the hypercapnia periods). Contrast changes are around $\pm2\%$

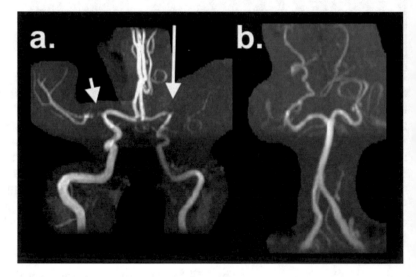

**Fig. 19** MR angiography of the head demonstrates the (**a**) abnormal anterior circulation with asymmetric flow enhancement being reduced in the left internal carotid artery and middle cerebral arteries bilaterally compared to the right internal carotid artery and anterior cerebral arteries due to occlusion of the left M1 segment with focal high-grade stenosis of the origin of the left A1 segment (long arrow) and high-grade focal stenosis of the right M1 segment (short arrow). The symmetric flow enhancement in the anterior cerebral arteries reflects the cross flow through the anterior communicating artery. (**b**) The posterior circulation shows symmetric flow enhancement in the posterior cerebral arteries which show expanded vascular territories

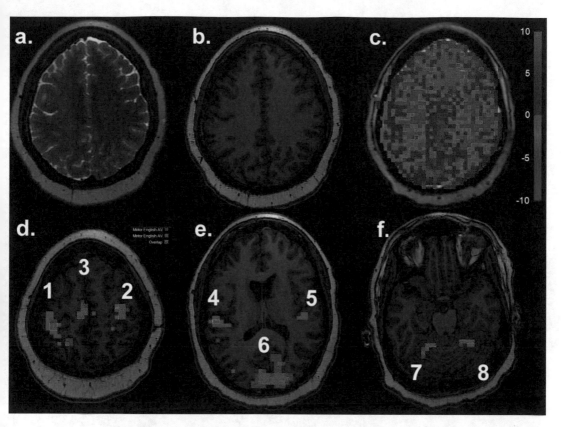

**Fig. 20** MR imaging demonstrates no anatomic abnormalities on (**a**) T2-weighted and (**b**) T1-weighted images but shows (**c**) loss of white matter vascular reactivity without loss of task-based BOLD contrast in (**d**) the primary sensorimotor (right 1, left 2) and supplementary motor areas (3), (**e**) auditory cortex (right 4, left 5) and visual cortex (6), and (**f**) cerebellum (right 7, left 8). It can be concluded that HR is still sufficient to support BOLD contrast in the cortex

(Fig. 22a) despite the stenosis of the extracranial left internal carotid artery with an incomplete circle of Willis with no visualized left posterior communicating artery (Fig. 22b, c). The vascular reactivity map (Fig. 22d) showed normal cortical vascular reactivity (Fig. 22d). The task-based BOLD response remained intact (Fig. 22e, f) indicating sufficient HR to support both BOLD contrast and vascular reactivity. Continued medical management is being pursued.

The QMRA results also indicated intact HR with flows in the left anterior (36–59 mL/min), middle (184–309 mL/min), and posterior (93–153 mL/min) cerebral arteries increasing significantly from baseline levels after vasodilation. The flow through the left posterior communicating artery also increased (118–162 mL/min) directed from the posterior to anterior circulation, indicating that the compensation in this case was at the circle of Willis as well as at the leptomeningeal level. The flow in the left A1 segment also reverses after vasodilation to supply flow from the

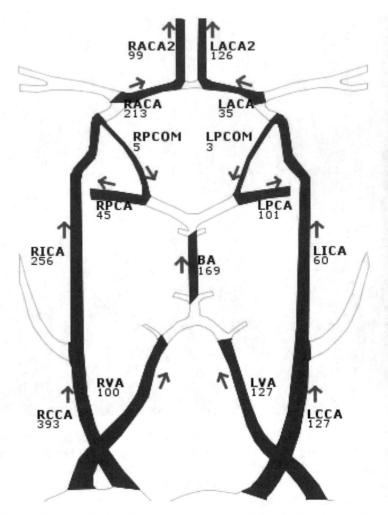

**Fig. 21** The quantitative flow measurements performed without vasodilation show the increased flow through the left anterior and posterior cerebral arteries supplying leptomeningeal flow into the compromised anterior circulation where the flows in the middle cerebral arteries were too low to measure accurately. The right anterior cerebral artery also shows increased flow for age consistent with its expanded vascular territory. Although the circle of Willis is intact, the posterior communicating arteries do not contribute significantly to the posterior circulation, apparently because of the expanded vascular territories of the anterior cerebral arteries. The lower flow in the left internal carotid artery was also consistent with the MRA examination shown in Fig. 19 and reflecting extracranial stenosis (not shown)

right internal carotid artery to the left side through the anterior communicating artery.

*4.1.4   Case D: Moyamoya Disease (Post-bypass Follow-Up)*

This 52-year-old right-handed woman with a medical history of hypertension, hypothyroidism, and seizures was found to have moyamoya disease. Initial evaluation of HR showed reduced

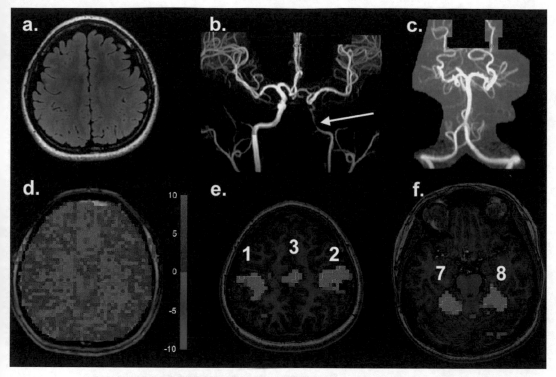

**Fig. 22** Although suspected of left internal carotid dissection after two episodes of spontaneously resolving dizziness and blurred vision with one resulting in loss of consciousness and fall, (**a**) T2-weighted FLAIR imaging and further comprehensive anatomic MR imaging showed no tissue injury. (**b**) MRA of the intracranial anterior circulation showed a stenotic left internal carotid artery not typical of dissection (arrow) and (**c**) a normal posterior circulation with an intact circle of Willis. (**d**) The vascular reactivity map was normal throughout the cortex with only symmetrically slightly decreased reactivity in the white matter. (**e**) Regional BOLD response from the task-based paradigm was intact in the primary sensorimotor (1, 2) and supplementary motor (3) areas and (**f**) control motor areas (7, 8) in the cerebellum. The auditory and visual areas were also intact (not shown)

vascular reactivity in the right middle and anterior cerebral artery territories, and she underwent right-sided STA-MCA bypass. She developed severe headaches following the procedure. She underwent follow-up re-evaluation of HR.

The MRA prior to surgical intervention in Fig. 23a showed the moyamoya configuration with occlusion of the right middle and bilateral anterior cerebral arteries proximally (arrow). The right posterior cerebral artery showed an expanded vascular territory consistent with leptomeningeal collateral flow (Fig. 23b). The immediate post-bypass MRA (Fig. 24) showed the right-sided STA-MCA bypass (arrows) in place to a prominent distal branch of the right middle cerebral artery. However, at the 6-month follow-up MRA (Fig. 25), the direct bypass was no longer patent (arrow), but the distal branches still showed flow enhancement similar to the preoperative study. Despite loss of the bypass, this

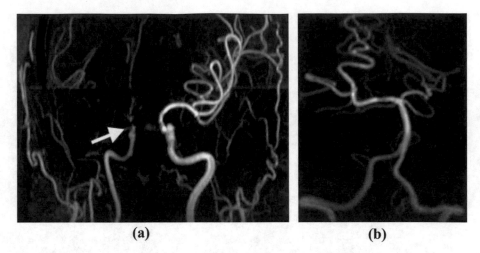

(a)                                    (b)

**Fig. 23** Prior to the right STA-MCA bypass, TOF MRA of (**a**) the anterior circulation and (**b**) posterior circulation showing occlusion (arrow) of the supraclinoid segment of the right internal carotid artery with loss of the A1 segments bilaterally and right M1 segment. The distal right middle cerebral artery branches are present but with reduced flow enhancement consistent with collateral flow. The vascular territory of the right posterior cerebral artery is expanded consistent with leptomeningeal collateral flow

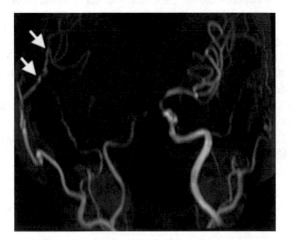

**Fig. 24** Anterior circulation of TOF MRA at 4 days following bypass surgery showing flow enhancement through the patent right STA-MCA bypass (arrows) into the distal branch of the right middle cerebral artery. The anterior cerebral arteries and proximal right middle cerebral artery show no flow enhancement indicating slow flow

flow is presumed to be due to the continued development of EDAS collateral flow. Catheter cerebral angiography also showed improving EDAS flow (not shown).

The fMRI protocol was used to evaluate the changing HR across the postoperative period as illustrated in Figs. 26 and 27.

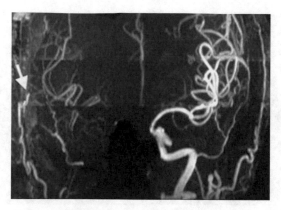

**Fig. 25** Follow-up TOF MRA at 6 months shows loss of the flow enhancement through the right STA-MCA bypass graft with abrupt attenuation (arrow) of flow enhancement in the right STA. However, flow enhancement is still present in the distal middle cerebral artery branches. No further ischemic changes were identified anatomically by MRI examination, and the patient remained asymptomatic. The right internal carotid artery was segmented from this maximum intensity projection but is unchanged from the previous study in Fig. 24

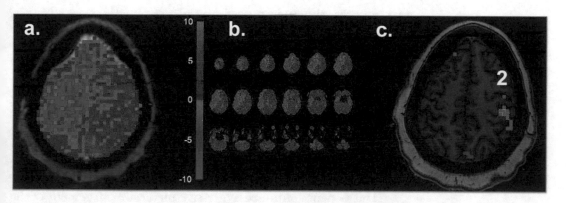

**Fig. 26** The complaint of worsening headaches after the right STA-MCA bypass for moyamoya disease and the occlusion of the bypass at 6-month follow-up led to re-evaluation of HR. No new anatomic changes had occurred since the pre-intervention study, but (**a**) the single axial partition from (**b**) the matrix of axial vascular reactivity maps showed reduced reactivity (blue) in the vascular territories of the right anterior and middle cerebral arteries. (**c**) The task-based BOLD activation was intact in the left primary sensorimotor (2) but not in the right sensorimotor cortex, bilateral supplementary motor cortex, or right primary auditory cortex although bilateral activation was present in the cerebellum that ensured that the patient was performing the task (not shown). The visual cortex was active bilaterally in the occipital lobes (not shown)

At the 6-month follow-up when the right STA-MCA bypass was no longer patent, vascular reactivity remained asymmetric and reduced in the right middle cerebral artery vascular territory (Fig. 26a, b) but without any anatomic changes and insufficient HR to support symmetric BOLD activation (Fig. 26c). Continued follow-up at

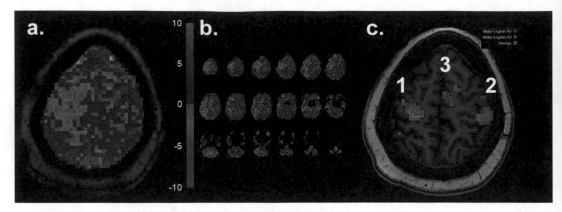

**Fig. 27** Ongoing evaluation of HR was performed at 16 months after the right STA-MCA bypass and 10 months after occlusion of the bypass was detected. No new anatomic changes had occurred since the pre-intervention study, but (**a**) the single axial partition from (**b**) the matrix of axial vascular reactivity maps still showed asymmetric reactivity (blue) but much less severe than at 6 months (Fig. 26a, b). (**c**) The task-based BOLD activation was now intact bilaterally in the primary sensorimotor cortex (1, 2) and left supplementary area (3). The bilateral auditory and visual areas were intact (not shown). Bilateral motor activation was present in the cerebellum that ensured that the patient was performing the task. This improvement in HR presumably reflects ongoing improved EDAS collateral flow as reflected on cerebral angiography (not shown)

16 months shows improved vascular reactivity (Fig. 27a, b) although still asymmetric and now with sufficient HR to support symmetric BOLD activation in the vascular territories of the right middle and left anterior cerebral arteries (Fig. 27c), presumably secondary to continued development of additional EDAS vessels.

QMRA measurements in the right posterior cerebral and basilar arteries were elevated for age prior to bypass, decreasing in the immediate postoperative period but increasing again when the bypass occluded but then decreasing again as the EDAS vessels continued to develop, presumably indicating decreasing need for leptomeningeal flow.

This patient continued aggressive medical management of the postprocedural headaches and will continue to be monitored for changes in HR using this QMRA and fMRI protocol annually, along with less frequent cerebral angiograms.

*4.2    Summary*

The integrity of HR is an important parameter for the management of cerebrovascular disease. We have highlighted the MR imaging methods used in our clinical service that we have found valuable and complementary to catheter cerebral angiography and computed tomography (CT) angiography that allow rational management of an evolving clinical history of cerebral ischemia in chronic steno-occlusive disease to limit future ischemic events.

Although neither loss of a regional BOLD fMRI response nor compromise of vascular reactivity detected using BOLD contrast during a hypercapnia challenge has been established to predict stroke in a prospective clinical trial, both indicate reduced HR and compromised vascular physiology. Our clinical experience for over the last 10 years using these approaches provides empirical evidence that these MR methods can be applied routinely in a clinical service to help manage intracranial disease.

## Acknowledgments

KRT acknowledges financial support from NIH RO1 NS059745 stroke.

*Disclosures:* KRT is the owner of the commercially available MRIx Technologies software used for the analysis of fMRI studies in this chapter.

## References

1. Aichner F (1994) Magnetic resonance angiography in cerebrovascular disease: a clinical update. J Stroke Cerebrovasc Dis 4(Suppl 1): S36–S40. https://doi.org/10.1016/s1052-3057(10)80253-8

2. Bryant DJ, Payne JA, Firmin DN, Longmore DB (1984) Measurement of flow with NMR imaging using a gradient pulse and phase difference technique. J Comput Assist Tomogr 8 (4):588–593

3. Moran PR (1982) A flow velocity zeugmatographic interlace for NMR imaging in humans. Magn Reson Imaging 1(4):197–203

4. Parker DL, Parker DJ, Blatter DD et al (1996) The effect of image resolution on vessel signal in high-resolution magnetic resonance angiography. J Magn Reson Imaging 6 (4):632–641

5. Pelc LR, Pelc NJ, Rayhill SC et al (1992) Arterial and venous blood flow: noninvasive quantitation with MR imaging. Radiology 185(3):809–812. https://doi.org/10.1148/radiology.185.3.1438767

6. Chatzimavroudis GP, Oshinski JN, Franch RH et al (2001) Evaluation of the precision of magnetic resonance phase velocity mapping for blood flow measurements. J Cardiovasc Magn Reson 3(1):11–19. https://doi.org/10.1081/jcmr-100000142

7. Khodarahmi I, Shakeri M, Kotys-Traughber M et al (2014) In vitro validation of flow measurement with phase contrast MRI at 3 tesla using stereoscopic particle image velocimetry and stereoscopic particle image velocimetry-based computational fluid dynamics. J Magn Reson Imaging 39 (6):1477–1485. https://doi.org/10.1002/jmri.24322

8. Ku DN, Biancheri CL, Pettigrew RI et al (1990) Evaluation of magnetic resonance velocimetry for steady flow. J Biomech Eng 112(4):464–472. https://doi.org/10.1115/1.2891212

9. Meier D, Maier S, Bosiger P (1988) Quantitative flow measurements on phantoms and on blood vessels with MR. Magn Reson Med 8(1):25–34

10. Walker MF, Souza SP, Dumoulin CL (1988) Quantitative flow measurement in phase contrast MR angiography. J Comput Assist Tomogr 12(2):304–313

11. Zhao M, Curcio AP, Clark ME et al (2004) In vitro validation of MR volumetric flow measurement. In: Proceedings of the 2004 International Workshop on Flow and Motion, pp 148–149

12. Firmin DN, Nayler GL, Klipstein RH et al (1987) In vivo validation of MR velocity imaging. J Comput Assist Tomogr 11 (5):751–756

13. Laffon E, Valli N, Latrabe V et al (1998) A validation of a flow quantification by MR phase mapping software. Eur J Radiol 27 (2):166–172

14. Maier SE, Meier D, Boesiger P et al (1989) Human abdominal aorta: comparative measurements of blood flow with MR imaging and multigated Doppler US. Radiology 171

(2):487–492. https://doi.org/10.1148/radi ology.171.2.2649924

15. Van Rossum AC, Sprenger M, Visser FC, Peels KH, Valk J, Roos JP (1991) An in vivo validation of quantitative blood flow imaging in arteries and veins using magnetic resonance phase-shift techniques. Eur Heart J 12 (2):117–126. https://doi.org/10.1093/oxfordjournals.eurheartj.a059857

16. Hofman MB, Visser FC, van Rossum AC et al (1995) In vivo validation of magnetic resonance blood volume flow measurements with limited spatial resolution in small vessels. Magn Reson Med 33(6):778–784

17. Wendt RE III, Rokey R, Wong WF, Marks A (1992) Magnetic resonance velocity measurements in small arteries. Comparison with Doppler ultrasonic measurements in the aortas of normal rabbits. Investig Radiol 27 (7):499–503

18. Bendel P, Buonocore E, Bockisch A, Besozzi MC (1989) Blood flow in the carotid arteries: quantification by using phase-sensitive MR imaging. AJR Am J Roentgenol 152 (6):1307–1310. https://doi.org/10.2214/ajr.152.6.1307

19. Buijs PC, Krabbe-Hartkamp MJ, Bakker CJ et al (1998) Effect of age on cerebral blood flow: measurement with ungated two-dimensional phase-contrast MR angiography in 250 adults. Radiology 209 (3):667–674. https://doi.org/10.1148/radi ology.209.3.9844657

20. Enzmann DR, Ross MR, Marks MP, Pelc NJ (1994) Blood flow in major cerebral arteries measured by phase-contrast cine MR. AJNR Am J Neuroradiol 15(1):123–129

21. Hendrikse J, van Raamt AF, van der Graaf Y et al (2005) Distribution of cerebral blood flow in the circle of Willis. Radiology 235 (1):184–189. https://doi.org/10.1148/radiol.2351031799

22. Marks MP, Pelc NJ, Ross MR, Enzmann DR (1992) Determination of cerebral blood flow with a phase-contrast cine MR imaging technique: evaluation of normal subjects and patients with arteriovenous malformations. Radiology 182(2):467–476. https://doi.org/10.1148/radiology.182.2.1732966

23. Mattle H, Edelman RR, Wentz KU et al (1991) Middle cerebral artery: determination of flow velocities with MR angiography. Radiology 181(2):527–530. https://doi.org/10.1148/radiology.181.2.1924799

24. Ravensbergen J, Tarnawski M, Vriens EM et al (1996) New ways of performing in vivo flow velocity measurements in the basilar artery. Neuroradiology 38(1):1–5

25. van Everdingen KJ, Klijn CJ, Kappelle LJ et al (1997) MRA flow quantification in patients with a symptomatic internal carotid artery occlusion. The Dutch EC-IC Bypass Study Group. Stroke 28(8):1595–1600. https://doi.org/10.1161/01.str.28.8.1595

26. Pelc NJ, Herfkens RJ, Shimakawa A, Enzmann DR (1991) Phase contrast cine magnetic resonance imaging. Magn Reson Q 7(4):229–254

27. Bakker CJ, Hartkamp MJ, Mali WP (1996) Measuring blood flow by nontriggered 2D phase-contrast MR angiography. Magn Reson Imaging 14(6):609–614

28. Bakker CJ, Kouwenhoven M, Hartkamp MJ et al (1995) Accuracy and precision of time-averaged flow as measured by nontriggered 2D phase-contrast MR angiography, a phantom evaluation. Magn Reson Imaging 13 (7):959–965

29. Enzmann DR, Marks MP, Pelc NJ (1993) Comparison of cerebral artery blood flow measurements with gated cine and ungated phase-contrast techniques. J Magn Reson Imaging 3(5):705–712

30. Maier SE, Cline HE, Jolesz FA (1995) Estimation of average flow in ungated 3D phase contrast angiograms. Magn Reson Med 34 (5):706–712

31. Zhao M, Charbel FT, Alperin N et al (2000) Improved phase-contrast flow quantification by three-dimensional vessel localization. Magn Reson Imaging 18(6):697–706

32. Lorensen WE, Cline HE (1987) Marching cubes: a high resolution 3D surface construction algorithm. ACM SIGGRAPH Comp Graph 21(4):163–169

33. Ruland S, Zhao M, Pandey D et al (2006) Reproducibility of cerebral blood flow analysis using quantitative magnetic resonance angiography. In: AANS/CNS Cerebrovascular Section 9th Joint Annual Meeting, February 17–20, 2006, Orlando, FL

34. Zhao M, Ruland S, Pandey D et al (2006) Repeatability of MR volumetric flow measurements in major cerebral arteries. In: 2006 ISMRM Flow and Motion Study Group Workshop: Imaging Assessment of Cardiovascular and Tissue Mechanics, July 13–16, 2006, New York, NY

35. Vorstrup S, Brun B, Lassen NA (1986) Evaluation of the cerebral vasodilatory capacity by the acetazolamide test before EC-IC bypass surgery in patients with occlusion of the

internal carotid artery. Stroke 17 (6):1291–1298. https://doi.org/10.1161/01.str.17.6.1291

36. Yonas H, Pindzola RR (1994) Physiological determination of cerebrovascular reserves and its use in clinical management. Cerebrovasc Brain Metab Rev 6(4):325–340

37. Caputi L, Ghielmetti F, Farago G et al (2014) Cerebrovascular reactivity by quantitative magnetic resonance angiography with a Co (2) challenge. Validation as a new imaging biomarker. Eur J Radiol 83(6):1005–1010. https://doi.org/10.1016/j.ejrad.2014.03.001

38. de Boorder MJ, Hendrikse J, van der Grond J (2004) Phase-contrast magnetic resonance imaging measurements of cerebral autoregulation with a breath-hold challenge: a feasibility study. Stroke 35(6):1350–1354. https://doi.org/10.1161/01.str.0000128530.75424.63

39. Vagal AS, Leach JL, Fernandez-Ulloa M, Zuccarello M (2009) The acetazolamide challenge: techniques and applications in the evaluation of chronic cerebral ischemia. AJNR Am J Neuroradiol 30(5):876–884. https://doi.org/10.3174/ajnr.A1538

40. Bahr-Hosseini M, Shakur SF, Amin-Hanjani-S, Charbel FT, Alaraj A (2016) Angiographic correlates of cerebral hemodynamic changes with diamox challenge assessed by quantitative magnetic resonance angiography. Stroke 47(6):1658–1660. https://doi.org/10.1161/strokeaha.116.013015

41. Calderon-Arnulphi M, Amin-Hanjani S, Alaraj A et al (2011) In vivo evaluation of quantitative MR angiography in a canine carotid artery stenosis model. AJNR Am J Neuroradiol 32(8):1552–1559. https://doi.org/10.3174/ajnr.A2546

42. Amin-Hanjani S, Du X, Pandey DK et al (2015) Effect of age and vascular anatomy on blood flow in major cerebral vessels. J Cereb Blood Flow Metab 35(2):312–318. https://doi.org/10.1038/jcbfm.2014.203

43. Zhao M, Amin-Hanjani S, Ruland S et al (2007) Regional cerebral blood flow using quantitative MR angiography. AJNR Am J Neuroradiol 28(8):1470–1473. https://doi.org/10.3174/ajnr.A0582

44. Correia de Verdier M, Wikstrom J (2016) Normal ranges and test-retest reproducibility of flow and velocity parameters in intracranial arteries measured with phase-contrast magnetic resonance imaging. Neuroradiology 58 (5):521–531. https://doi.org/10.1007/s00234-016-1661-6

45. Ross MR, Pelc NJ, Enzmann DR (1993) Qualitative phase contrast MRA in the normal and abnormal circle of Willis. AJNR Am J Neuroradiol 14(1):19–25

46. Hagstadius S, Risberg J (1989) Regional cerebral blood flow characteristics and variations with age in resting normal subjects. Brain Cogn 10(1):28–43

47. Brisman JL, Pile-Spellman J, Konstas AA (2012) Clinical utility of quantitative magnetic resonance angiography in the assessment of the underlying pathophysiology in a variety of cerebrovascular disorders. Eur J Radiol 81 (2):298–302. https://doi.org/10.1016/j.ejrad.2010.12.079

48. Holmstedt CA, Turan TN, Chimowitz MI (2013) Atherosclerotic intracranial arterial stenosis: risk factors, diagnosis, and treatment. Lancet Neurol 12(11):1106–1114. https://doi.org/10.1016/s1474-4422(13)70195-9

49. Bae YJ, Jung C, Kim JH et al (2015) Quantitative magnetic resonance angiography in internal carotid artery occlusion with primary collateral pathway. J Stroke 17(3):320–326. https://doi.org/10.5853/jos.2015.17.3.320

50. Ruland S, Ahmed A, Thomas K et al (2009) Leptomeningeal collateral volume flow assessed by quantitative magnetic resonance angiography in large-vessel cerebrovascular disease. J Neuroimaging 19(1):27–30. https://doi.org/10.1111/j.1552-6569.2008.00249.x

51. McVerry F, Liebeskind DS, Muir KW (2012) Systematic review of methods for assessing leptomeningeal collateral flow. AJNR Am J Neuroradiol 33(3):576–582. https://doi.org/10.3174/ajnr.A2794

52. Amin-Hanjani S, Du X, Zhao M et al (2005) Use of quantitative magnetic resonance angiography to stratify stroke risk in symptomatic vertebrobasilar disease. Stroke 36 (6):1140–1145. https://doi.org/10.1161/01.STR.0000166195.63276.7c

53. Amin-Hanjani S, Pandey DK, Rose-Finnell L et al (2016) Effect of hemodynamics on stroke risk in symptomatic atherosclerotic vertebrobasilar occlusive disease. JAMA Neurol 73(2):178–185. https://doi.org/10.1001/jamaneurol.2015.3772

54. Bauer AM, Amin-Hanjani S, Alaraj A, Charbel FT (2009) Quantitative magnetic resonance angiography in the evaluation of the subclavian steal syndrome: report of 5 patients. J Neuroimaging 19(3):250–252. https://doi.org/10.1111/j.1552-6569.2008.00297.x

55. Douglas AF, Christopher S, Amankulor N et al (2011) Extracranial carotid plaque length and parent vessel diameter significantly affect baseline ipsilateral intracranial blood flow. Neurosurgery 69(4):767–773.; discussion 773. https://doi.org/10.1227/NEU. 0b013e31821ff8f4

56. Ghogawala Z, Amin-Hanjani S, Curran J et al (2013) The effect of carotid endarterectomy on cerebral blood flow and cognitive function. J Stroke Cerebrovasc Dis 22 (7):1029–1037. https://doi.org/10.1016/j. jstrokecerebrovasdis.2012.03.016

57. Derdeyn CP, Chimowitz MI, Lynn MJ et al (2014) Aggressive medical treatment with or without stenting in high-risk patients with intracranial artery stenosis (SAMMPRIS): the final results of a randomised trial. Lancet 383(9914):333–341. https://doi.org/10. 1016/s0140-6736(13)62038-3

58. Zaidat OO, Castonguay AC, Fitzsimmons BF et al (2013) Design of the Vitesse Intracranial Stent Study for Ischemic Therapy (VISSIT) trial in symptomatic intracranial stenosis. J Stroke Cerebrovasc Dis 22(7):1131–1139. https://doi.org/10.1016/j. jstrokecerebrovasdis.2012.10.021

59. Guppy KH, Charbel FT, Corsten LA et al (2002) Hemodynamic evaluation of basilar and vertebral artery angioplasty. Neurosurgery 51(2):327–333. discussion 333-324

60. Langer DJ, Lefton DR, Ostergren L et al (2006) Hemispheric revascularization in the setting of carotid occlusion and subclavian steal: a diagnostic and management role for quantitative magnetic resonance angiography? Neurosurgery 58(3):528–533.; discussion 528-533. https://doi.org/10.1227/01. neu.0000197331.41985.15

61. Prabhakaran S, Wells KR, Jhaveri MD, Lopes DK (2011) Hemodynamic changes following wingspan stent placement--a quantitative magnetic resonance angiography study. J Neuroimaging 21(2):e109–e113. https:// doi.org/10.1111/j.1552-6569.2009. 00425.x

62. Shakur SF, Amin-Hanjani S, Bednarski C et al (2015) Intracranial blood flow changes after extracranial carotid artery stenting. Neurosurgery 76(3):330–336. https://doi.org/10. 1227/neu.0000000000000618

63. Amin-Hanjani S, Alaraj A, Calderon-Arnulphi M et al (2010) Detection of intracranial in-stent restenosis using quantitative magnetic resonance angiography. Stroke 41 (11):2534–2538. https://doi.org/10.1161/ strokeaha.110.594739

64. Prabhakaran S, Warrior L, Wells KR et al (2009) The utility of quantitative magnetic resonance angiography in the assessment of intracranial in-stent stenosis. Stroke 40 (3):991–993. https://doi.org/10.1161/ strokeaha.108.522391

65. Sun W, Ruan Z, Dai X et al (2018) Quantifying hemodynamic changes in moyamoya disease based on two-dimensional cine phase-contrast magnetic resonance imaging and computational fluid dynamics. World Neurosurg 120:e1301–e1309. https://doi.org/10. 1016/j.wneu.2018.09.057

66. Khan N, Lober RM, Ostergren L et al (2017) Measuring cerebral blood flow in moyamoya angiopathy by quantitative magnetic resonance angiography noninvasive optimal vessel analysis. Neurosurgery 81(6):921–927. https://doi.org/10.1093/neuros/nyw122

67. Neff KW, Horn P, Schmiedek P et al (2006) 2D cine phase-contrast MRI for volume flow evaluation of the brain-supplying circulation in moyamoya disease. AJR Am J Roentgenol 187(1):W107–W115. https://doi.org/10. 2214/ajr.05.0219

68. Amin-Hanjani S, Shin JH, Zhao M et al (2007) Evaluation of extracranial-intracranial bypass using quantitative magnetic resonance angiography. J Neurosurg 106(2):291–298. https://doi.org/10.3171/jns.2007.106.2. 291

69. Kim T, Bang JS, Kwon OK et al (2017) Hemodynamic changes after unilateral revascularization for moyamoya disease: serial assessment by quantitative magnetic resonance angiography. Neurosurgery 81 (1):111–119. https://doi.org/10.1093/neu ros/nyw035

70. Zhu F, Qian Y, Xu B et al (2018) Quantitative assessment of changes in hemodynamics of the internal carotid artery after bypass surgery for moyamoya disease. J Neurosurg 129 (3):677–683. https://doi.org/10.3171/ 2017.5.jns163112

71. Amin-Hanjani S, Singh A, Rifai H et al (2013) Combined direct and indirect bypass for moyamoya: quantitative assessment of direct bypass flow over time. Neurosurgery 73 (6):962–967.; discussion 967-968. https:// doi.org/10.1227/neu.0000000000000139

72. Krishnamurthy R, Bahouth SM, Muthupillai R (2016) 4D Contrast-enhanced MR angiography with the keyhole technique in children: technique and clinical applications. Radiographics 36(2):523–537. https://doi.org/ 10.1148/rg.2016150106

73. Willinek WA, Hadizadeh DR, von Falkenhausen M et al (2008) 4D time-resolved MR

angiography with keyhole (4D-TRAK): more than 60 times accelerated MRA using a combination of CENTRA, keyhole, and SENSE at 3.0T. J Magn Reson Imaging 27 (6):1455–1460. https://doi.org/10.1002/jmri.21354

74. Meng L, Gelb AW (2015) Regulation of cerebral autoregulation by carbon dioxide. Anesthesiology 122(1):196–205. https://doi.org/10.1097/aln.0000000000000506

75. Peterson EC, Wang Z, Britz G (2011) Regulation of cerebral blood flow. Int J Vasc Med 2011:823525. https://doi.org/10.1155/2011/823525

76. Strandgaard S, Olesen J, Skinhoj E, Lassen NA (1973) Autoregulation of brain circulation in severe arterial hypertension. Br Med J 1(5852):507–510. https://doi.org/10.1136/bmj.1.5852.507

77. Berg JM, Tymoczko JL, Stryer L (2002) Hemoglobin transports oxygen efficiently by binding oxygen cooperatively. In: Biochemistry, 5th edn. W H Freeman, New York, NY

78. Cipolla MJ (2009) Anatomy and ultrastructure. In: The cerebral circulation. Morgan & Claypool Life Sciences, San Rafael, CA

79. Duvernoy HM, Delon S, Vannson JL (1981) Cortical blood vessels of the human brain. Brain Res Bull 7(5):519–579. https://doi.org/10.1016/0361-9230(81)90007-1

80. Clarke DD, Sokoloff L (1999) Regulation of cerebral metabolic rate. In: Siegel GJ, Agranoff BW, Albers RW et al (eds) Basic neurochemistry: molecular, cellular and medical aspects, 6th edn. Lippincott-Raven, Philadelphia, PA

81. Karbowski J (2011) Scaling of brain metabolism and blood flow in relation to capillary and neural scaling. PLoS One 6(10):e26709. https://doi.org/10.1371/journal.pone.0026709

82. Ogawa S, Tank DW, Menon R et al (1992) Intrinsic signal changes accompanying sensory stimulation: functional brain mapping with magnetic resonance imaging. Proc Natl Acad Sci U S A 89(13):5951–5955. https://doi.org/10.1073/pnas.89.13.5951

83. Ogawa S, Menon RS, Tank DW et al (1993) Functional brain mapping by blood oxygenation level-dependent contrast magnetic resonance imaging. A comparison of signal characteristics with a biophysical model. Biophys J 64(3):803–812. https://doi.org/10.1016/s0006-3495(93)81441-3

84. Bandettini PA, Wong EC, Hinks RS et al (1992) Time course EPI of human brain function during task activation. Magn Reson Med 25(2):390–397. https://doi.org/10.1002/mrm.1910250220

85. Kwong KK, Belliveau JW, Chesler DA et al (1992) Dynamic magnetic resonance imaging of human brain activity during primary sensory stimulation. Proc Natl Acad Sci U S A 89 (12):5675–5679. https://doi.org/10.1073/pnas.89.12.5675

86. Gauthier CJ, Fan AP (2019) BOLD signal physiology: models and applications. NeuroImage 187:116–127. https://doi.org/10.1016/j.neuroimage.2018.03.018

87. Herscovitch P, Mintun MA, Raichle ME (1985) Brain oxygen utilization measured with oxygen-15 radiotracers and positron emission tomography: generation of metabolic images. J Nucl Med 26(4):416–417

88. Pantano P, Baron JC, Lebrun-Grandie P et al (1984) Regional cerebral blood flow and oxygen consumption in human aging. Stroke 15 (4):635–641. https://doi.org/10.1161/01.str.15.4.635

89. Ter-Pogossian MM, Herscovitch P (1985) Radioactive oxygen-15 in the study of cerebral blood flow, blood volume, and oxygen metabolism. Semin Nucl Med 15 (4):377–394

90. Powers WJ, Press GA, Grubb RL et al (1987) The effect of hemodynamically significant carotid artery disease on the hemodynamic status of the cerebral circulation. Ann Intern Med 106(1):27–34. https://doi.org/10.7326/0003-4819-106-1-27

91. Sette G, Baron JC, Mazoyer B et al (1989) Local brain haemodynamics and oxygen metabolism in cerebrovascular disease. Positron emission tomography. Brain 112 (Pt 4):931–951. https://doi.org/10.1093/brain/112.4.931

92. Powers WJ (1991) Cerebral hemodynamics in ischemic cerebrovascular disease. Ann Neurol 29(3):231–240. https://doi.org/10.1002/ana.410290302

93. Ito H, Kanno I, Fukuda H (2005) Human cerebral circulation: positron emission tomography studies. Ann Nucl Med 19 (2):65–74

94. Kamath A, Smith WS, Powers WJ et al (2008) Perfusion CT compared to H(2) (15)O/O (15)O PET in patients with chronic cervical carotid artery occlusion. Neuroradiology 50 (9):745–751. https://doi.org/10.1007/s00234-008-0403-9

95. Derdeyn CP, Videen TO, Yundt KD et al (2002) Variability of cerebral blood volume and oxygen extraction: stages of cerebral haemodynamic impairment revisited. Brain 125

(Pt 3):595–607. https://doi.org/10.1093/brain/awf047

96. Aoe J, Watabe T, Shimosegawa E et al (2018) Evaluation of the default-mode network by quantitative (15)O-PET: comparative study between cerebral blood flow and oxygen consumption. Ann Nucl Med 32(7):485–491. https://doi.org/10.1007/s12149-018-1272-x

97. Guadagno JV, Jones PS, Fryer TD et al (2006) Local relationships between restricted water diffusion and oxygen consumption in the ischemic human brain. Stroke 37 (7):1741–1748. https://doi.org/10.1161/01.str.0000232437.00621.86

98. Fierstra J, van Niftrik C, Warnock G et al (2018) Staging hemodynamic failure with blood oxygen-level-dependent functional magnetic resonance imaging cerebrovascular reactivity: a comparison versus gold standard ((15)O-)H2O-positron emission tomography. Stroke 49(3):621–629. https://doi.org/10.1161/strokeaha.117.020010

99. Huettel SA, Song AW, McCarthy G (2009) Functional magnetic resonance imaging, 2nd edn. Oxford University Press, Oxford

100. Thulborn KR, Waterton JC, Matthews PM, Radda GK (1982) Oxygenation dependence of the transverse relaxation time of water protons in whole blood at high field. Biochim Biophys Acta 714(2):265–270. https://doi.org/10.1016/0304-4165(82)90333-6

101. Thulborn KR, Atkinson IC, Alexander A et al (2018) Comparison of blood oxygenation level-dependent fMRI and provocative DSC perfusion MR imaging for monitoring cerebrovascular reserve in intracranial chronic cerebrovascular disease. AJNR Am J Neuroradiol 39(3):448–453. https://doi.org/10.3174/ajnr.A5515

102. Thulborn KR (1999) Clinical rationale for very-high-field (3.0 Tesla) functional magnetic resonance imaging. Top Magn Reson Imaging 10(1):37–50

103. Thulborn KR (1999) Visual feedback to stabilize head position for fMRI. Magn Reson Med 41(5):1039–1043. https://doi.org/10.1002/(sici)1522-2594(199905)41:5<1039::aid-mrm24>3.0.co;2-n

104. De Vis JB, Bhogal AA, Hendrikse J et al (2018) Effect sizes of BOLD CVR, resting-state signal fluctuations and time delay measures for the assessment of hemodynamic impairment in carotid occlusion patients. NeuroImage 179:530–539. https://doi.org/10.1016/j.neuroimage.2018.06.017

105. Juttukonda MR, Donahue MJ (2019) Neuroimaging of vascular reserve in patients with cerebrovascular diseases. NeuroImage 187:192–208. https://doi.org/10.1016/j.neuroimage.2017.10.015

106. Mandell DM, Han JS, Poublanc J et al (2008) Mapping cerebrovascular reactivity using blood oxygen level-dependent MRI in Patients with arterial steno-occlusive disease: comparison with arterial spin labeling MRI. Stroke 39(7):2021–2028. https://doi.org/10.1161/strokeaha.107.506709

107. Poublanc J, Han JS, Mandell DM et al (2013) Vascular steal explains early paradoxical blood oxygen level-dependent cerebrovascular response in brain regions with delayed arterial transit times. Cerebrovasc Dis Extra 3 (1):55–64. https://doi.org/10.1159/000348841

108. Poublanc J, Crawley AP, Sobczyk O et al (2015) Measuring cerebrovascular reactivity: the dynamic response to a step hypercapnic stimulus. J Cereb Blood Flow Metab 35 (11):1746–1756. https://doi.org/10.1038/jcbfm.2015.114

109. Sobczyk O, Battisti-Charbonney A, Poublanc J et al (2015) Assessing cerebrovascular reactivity abnormality by comparison to a reference atlas. J Cereb Blood Flow Metab 35 (2):213–220. https://doi.org/10.1038/jcbfm.2014.184

110. Spano VR, Mandell DM, Poublanc J et al (2013) CO2 blood oxygen level-dependent MR mapping of cerebrovascular reserve in a clinical population: safety, tolerability, and technical feasibility. Radiology 266 (2):592–598. https://doi.org/10.1148/radiol.12112795

# Chapter 9

# Breath-Hold Cerebrovascular Reactivity Mapping for Neurovascular Uncoupling Assessment in Primary Gliomas

## Domenico Zacà, Shruti Agarwal, and Jay J. Pillai

## Abstract

BOLD functional MRI (fMRI) for presurgical planning allows maximal brain tumor surgical resection with preservation of cortical areas critical for sensory and cognitive function. This technique can thus help to dramatically extend survival while maintaining good quality of life in these patients. Primary gliomas can alter the relationship between neuronal metabolic activity and vascular response which is the basis of BOLD fMRI. This phenomenon, reported as neurovascular uncoupling (NVU), may result in false-negative or abnormally reduced activation potentially leading neurosurgeons to remove cortical regions erroneously deemed as non-eloquent. Breath-hold cerebrovascular reactivity (BH CVR) mapping can provide a clinically viable and reliable method to assess NVU potential in both low- and high-grade gliomas. In this chapter, we will first explain BH CVR methodology, both in terms of acquisition and analysis. Then we will show how BH CVR can help in assessing NVU potential that can adversely affect interpretation of both task-based activation and resting-state functional connectivity maps. Finally, we will briefly mention alternative resting-state fMRI-based methods of NVU detection. BH CVR mapping represents an essential quality control tool that should be routinely used in clinical fMRI of primary gliomas.

**Keywords** Functional MRI, Gliomas, Neurovascular uncoupling, Cerebrovascular reactivity, Resting-state fMRI

## 1 Introduction

Blood oxygen level-dependent functional magnetic resonance imaging (BOLD fMRI) is a technique that is able to map brain functional activation by exploiting the transient change in regional oxygen extraction fraction accompanying neural activation [1].

The requirement of increased oxygen supply by the activated areas is coupled with a much larger increase in cerebral blood flow (CBF) and cerebral blood volume (CBV). This causes a decrease in regional deoxyhemoglobin concentration which results in an increase in the magnetic relaxation times T2 and T2*.

Jean Chen and Jorn Fierstra (eds.), *Cerebrovascular Reactivity: Methodological Advances and Clinical Applications*, Neuromethods, vol. 175, https://doi.org/10.1007/978-1-0716-1763-2_9, © Springer Science+Business Media, LLC, part of Springer Nature 2022

By acquiring T2- or T2*-weighted images while a subject is performing a task, it is possible to map the activated regions of the brain by detecting the areas where a statistically significant increase of the BOLD signal relative to a baseline has occurred.

A gradient echo sequence with echo planar readout (EPI) is routinely used for BOLD fMRI because it provides excellent sensitivity to the T2* effect [2]. At the same time, it allows to scan the whole brain with sufficient temporal resolution (1–2 s) and spatial resolution (2–3 mm) to adequately detect the BOLD signal increase related to functional activation.

BOLD fMRI was introduced in the early 1990s by Ogawa and colleagues [3]. Widely used since then as a brain mapping technique in cognitive neurosciences, BOLD fMRI has slowly made its transition from purely research to a clinically viable technique for presurgical mapping of eloquent cortical areas in patients with brain tumor and other potentially resectable lesions (e.g., epilepsy and vascular malformations) [4].

## 2   Clinical BOLD fMRI

The goal of clinical BOLD fMRI is to identify cortical eloquent areas at risk of being resected during surgical removal of focal lesions. As, in brain tumors, the extent of resection is one of the most critical prognostic factors affecting patient survival, the localization of sensorimotor, language, and visual cortical areas in spatial proximity of a lesion is of paramount importance in order to plan a safe while effective surgical removal of a tumor, an epilepsy focus, or a vascular malformation [5].

Currently, the protocols for presurgical fMRI recommend the use of block-designed paradigms alternating with periodicity blocks of rest and active tasks for a total duration of 3–4 min [6]. A prescan training session is also strongly advised in order to assess patient's ability to perform a task.

Following data acquisition and multiple preprocessing steps, fMRI activation maps are generated using a general linear model (GLM) analysis. This analysis requires the use of a BOLD hemodynamic response model to which each preprocessed voxel time series is fit. The fit returns for each voxel a statistical score ($T$ or $Z$) with its statistical significance. All voxels whose score is above a predetermined threshold (e.g., $Z > 4$) comprise the activation maps (Fig. 1).

The hemodynamic response model used in GLM analysis describes the theoretical temporal evolution of the BOLD signal in the activation areas due to the coupling between neuronal firing and increase in blood flow, blood volume, and oxygenation (neurovascular coupling) [7]. However, in many brain diseases including brain tumors, the neurovascular coupling is known to be altered

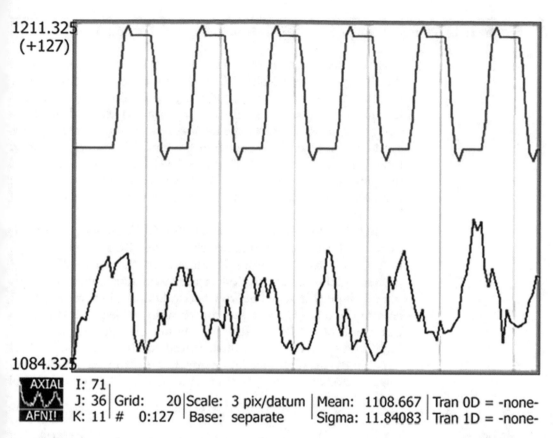

1211.325
(+127)

1084.325

| AXIAL | I: 71 | | | | | |
|---|---|---|---|---|---|---|
| AFNI! | J: 36 | Grid: 20 | Scale: 3 pix/datum | Mean: 1108.667 | Tran 0D = -none- |
| | K: 11 | # 0:127 | Base: separate | Sigma: 11.84083 | Tran 1D = -none- |

**Fig. 1** BOLD hemodynamic response (red line) and recorded (black line) BOLD signal in a voxel during a performance of a language task consisting of 20 s periods of rest alternating with a 20 s periods of a phonemic fluency task repeated 6 times. Very good correspondence can be noticed between the theoretical and experimental BOLD signal curve (i.e., high $Z$-score value) indicating this voxel belonging to an eloquent expressive language area

[8]. Several studies have reported decreased ipsilesional (i.e., hemisphere ipsilateral to the brain lesion) activation in comparison with contralesional activation in sensorimotor tasks as well as incorrect assessment of language dominance [9, 10]. This phenomenon referred to as neurovascular uncoupling (NVU) can potentially increase the risk of false negatives in BOLD fMRI presurgical mapping. Active cortical areas surrounding a lesion may go undetected and be erroneously deemed as non-eloquent if NVU is not properly taken into account. This could increase the risk of postoperative deficits due to inadvertent surgical resection of functional cortex. The main reason for such NVU detection is to alert the neurosurgeon to the need for complementary electrophysiologic (intraoperative cortical) mapping for more accurate delineation of eloquent cortex. Examples of other disease potentially affected by NVU include carotid occlusion, transient global ischemia, penumbra of cerebral ischemia, subarachnoid hemorrhage, epilepsy, and Alzheimer's disease (AD) [11–13].

## 3    Breath-Hold Cerebrovascular Reactivity Mapping for NVU Detection

Cerebrovascular reactivity (CVR) mapping is a technique to assess the change in cerebral blood flow in response to a vasodilatory or vasoconstrictive stimulus. Blood $CO_2$ partial pressure (pCO2) is thought to be responsible for maintaining the blood flow in the microcirculation distal to large feeding vessels, and therefore, CVR can be investigated by manipulating the pCO2 level in the blood [14]. Different imaging techniques have been applied for CVR mapping such as transcranial Doppler, single-photon emission computed tomography, positron emission tomography, and MRI [15, 16].

BOLD fMRI itself can be used for CVR mapping. As mentioned in the introduction of this chapter, the BOLD signal change in response to a stimulus is caused by the change in de-/oxyhemoglobin concentration in the vasculature adjacent to the site of stimulus-induced neuronal activation. An increase of pCO2 level in the brain blood vessels in a healthy subject will trigger an increase in blood flow without (or minimally) altering the metabolic activity levels. This will then lead to an increase of BOLD signal in the arterioles feeding eloquent cortex in the whole brain (Fig. 2). For this reason, BOLD CVR mapping can serve as a tool

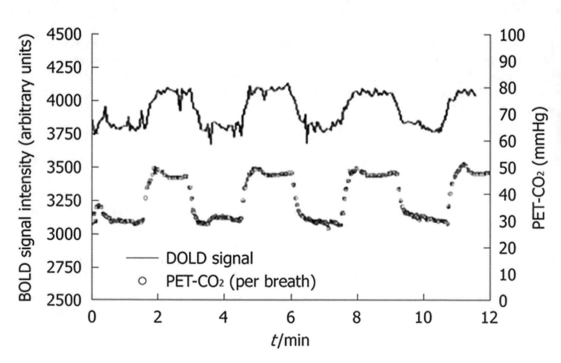

**Fig. 2** BOLD cerebrovascular reactivity measurement using administration of air with altered $CO_2$ or $O_2$ levels: the blood oxygen level-dependent (BOLD) signal pattern closely follows pressure of end-tidal $CO_2$ (end-tidal pressure of carbon dioxide) pattern (from Vesely et al). BOLD signal and pressure of end-tidal $CO_2$ (PET-$CO_2$) signals are reported in arbitrary units

to detect areas of NVU potential. Indeed, a cortical region characterized by reduced or absent CVR will not show BOLD signal increase in response to a neuronal stimulus expected to activate that region [17].

The increase of pCO2 level in the blood can be obtained with two approaches: a breath-hold (BH) task and the controlled inhalation of $CO_2$-manipulated air [18]. BH CVR is usually performed with a block design task alternating periods of BH with periods of normal respiration. Instructions to patients are sent by visual or auditory cues. CVR maps can be then computed and displayed as BOLD% signal change of the BH task response. Recently, it has been shown that the BOLD signal temporal lag and coherence maps combined with BOLD% signal change can provide a more robust and independent BH CVR assessment that is also independent of task compliance [19]. BH blocks of 15–20 s followed by 30–40 s of normal respiration blocks repeated 4–5 times have been shown to provide reproducible results and can be tolerated and adequately performed by most patients [20]. The main disadvantage of using BH CVR is the lack of quantitative measures that can be obtained. Nevertheless, since no special equipment is necessary, it can be easily added as an extra task in a clinical fMRI session. BH following inspiration (i.e., end-inspiratory BH) is easier to perform, although studies have shown BH following expiration is more repeatable [21].

CVR using manipulated $CO_2$ level can precisely control the content of inhaled air. The air and end tidal (ET) pCO$_2$ can be monitored during the experiments, and quantitative CVR maps can be calculated as the ratio between MR signal change due to $CO_2$ inhalation and the related pCO2 changes and expressed in units of percentage MR signal change/mmHg $CO_2$ [22]. The main advantage of these techniques over BH CVR is the quantitative information on CVR they can provide in comparison with BH. Quantitative CVR studies have been published in the literature applying and showing its utility in different cerebrovascular pathologies, such as moyamoya and steno-occlusive diseases [23]. However, despite recent progresses, the devices needed for this test require specific setup and patient training. It is then quite difficult to routinely apply this technique in a clinical MRI scenario. For this reason, to our knowledge, no $CO_2$ level-controlled (via exogenous gas administration) CVR studies have been published describing application in brain tumor presurgical mapping protocols.

## 4    BH CVR in Brain Tumor

In patients with focal resectable brain lesions such as brain tumors, epileptogenic tissue, or vascular malformations, NVU can potentially alter the BOLD signal response following neuronal activation [24]. This is caused by the disruption at the cellular, vascular, or metabolic level of the complex, and not completely yet understood, interplay of events at the microstructural, biochemical, and electrophysiologic levels, commonly referred to as neurovascular cascade, triggered by neuronal activation.

Therefore, when BOLD fMRI studies are carried out in these patients, BOLD activation detection may be suboptimal especially when data analysis models built upon numerous experiments performed in healthy volunteers are applied.

NVU is a recognized and thoroughly described phenomenon in brain tumors. High-grade gliomas (HGG) are characterized by tumor angiogenesis that is associated with abnormal vasoactivity and permeability of the neovasculature [25]. Indeed, since the early times of presurgical mapping by BOLD fMRI, there had been reports of HGG patients demonstrating decreased BOLD activation in the tumor adjacent to motor cortical areas in comparison with the contralesional cortical regions, which had been attributed to NVU [26]. The inverse correlation between BOLD activation strength and perfusion parameters such as cerebral blood volume and cerebral blood flow implies that in HGG patients, the disruption of the neurovascular cascade occurs at the neurovascular level [27]. The first studies of BH CVR mapping in these patients demonstrated a very high prevalence of reduced CVR in the tumor ipsilesional hemisphere associated with decreased BOLD activation and increased perfusion. These results thus showed the potential of BH CVR as a surrogate marker of NVU in HGG.

Low-grade gliomas (LGG) have been also shown to be affected in some cases by NVU similar to HGG [28]. As tumor angiogenesis is rarely present in LGG, the disruption of the neurovascular cascade may be due to astrocytic dysfunction or other yet unknown factors related to the infiltrative tumors [28]. For this reason, BH CVR mapping is particularly relevant as a technique for detection of NVU potential in LGG because perfusion imaging itself may not demonstrate alterations explaining decreased or absent activation in close spatial proximity to the tumor [29]. From the series reported in the literature, it also appears that NVU may not necessarily be an all-or-none (i.e., binary) phenomenon but can result in variable degrees of reduction of expected ipsilesional BOLD activation in eloquent cortical regions and complete absence of detectable activation only in some cases.

Similar to brain tumors, NVU regional hemodynamic abnormalities related to the exhaustion of vascular reserve and loss of perfusion pressure are likely to cause NVU in arteriovenous malformations [30].

## 5    BH CVR in Presurgical Mapping: Clinical Utility

In this section, we are going to illustrate two glioma cases, one high grade and one low grade, where BH CVR mapping results

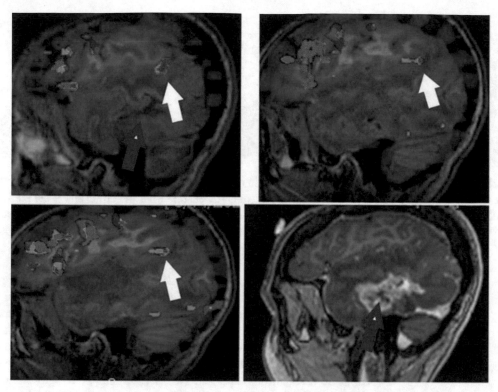

**Fig. 3** Composite language maps (see main text for the color legend of each task) and BH CVR map (low opacity blue) in this 45-year-old female patient undergoing presurgical fMRI before removal of a high-grade glioma in the left temporal lobe. The minimal activation below the superior temporal gyrus together with the decreased CVR (white arrow) suggested the presence of NVU in this area. The lower right image shows central necrosis in the left middle and inferior temporal gyrus (red arrow)

evaluated together with task fMRI activation pattern and patient's conditions clearly suggested NVU and, accordingly, influenced surgical planning.

*Case 1*: This 45-year-old female patient presented with recurrent IDH1 wild-type, MGMT-unmethylated WHO grade IV glioblastoma involving the left temporal lobe approximately 8 months following initial diagnosis. Notice the centrally necrotic peripherally irregularly enhancing mass with substantial surrounding T2/FLAIR hyperintense nonenhancing edema, indicative of nonenhancing tumor infiltration (the red arrow in the left upper quadrant image of Fig. 3 depicts the very low T1 signal area within the left middle [MTG] and inferior temporal gyri [ITG] corresponding to areas of central necrosis seen on the postcontrast image in the right lower quadrant). On the leakage-corrected relative cerebral blood volume perfusion map, note the high perfusion along the enhancing periphery of the mass, consistent with the high grade nature of the tumor, as shown by the white arrow in Fig. 4, first image on the left.

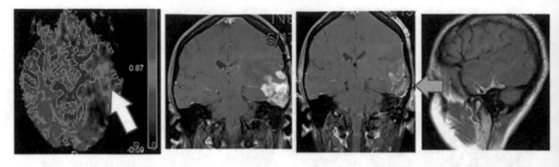

**Fig. 4** High perfusion (first from the left) can be seen in this 45-year-old female patient who underwent gross total resection of a left temporal lobe high-grade glioma. The second from the left image show the tumor appearance preoperatively on coronal postcontrast T1-weighted images. The third and fourth images from the left are postoperative T1-weighted images showing the surgical approach that was taken based on the results of preoperative fMRI that left the superior temporal gyrus intact

The composite language activation map displays seven different language tasks, each individually color-coded and thresholded (at $T$-values varying from approximately 3.7–4.5): green refers to an object naming task, purple refers to a passive story listening task, orange depicts a sentence listening comprehension task, light blue/cyan indicates the rhyming task, magenta indicates a noun-verb semantic association task, red refers to a silent word generation task, and yellow denotes a sentence completion task. Superimposed on this composite language activation map is a breath-hold cerebrovascular reactivity (BH CVR) map, which has been thresholded at 0.50 BOLD% signal change and overlaid using 30% opacity in dark blue. Please note that all displayed images follow radiologic convention, whereby the anatomic right cerebral hemisphere is shown in the left lateral aspect of the image and vice versa.

Notice the highly convergent Broca's area (BA) clusters of activation anteriorly in the left inferior frontal gyrus, as well as overall paucity of activation in the dominant left temporal lobe due to pathologically reduced regional CVR affecting the left superior temporal (STG) and middle temporal gyri (MTG), as depicted in the upper row images and left lower quadrant image. Although a small cluster of convergent activation is seen along the posteriormost aspect of the left STG, extending into the left supramarginal gyrus (SMG), as shown by the white arrows, no activation is seen more anteriorly or inferiorly within the STG, MTG, or inferior temporal gyrus as expected on these multiple language tasks. This paucity of activation in the left temporal lobe and regional impairment of CVR are indicative of tumor-induced neurovascular uncoupling (NVU), since the patient did not present with a Wernicke's (receptive) aphasia and was able to perform all language tasks without difficulty.

Postoperative images show gross total resection via a lateral inferior approach with the left STG left intact. The second image from the left in Fig. 4 represents a coronal postcontrast T1-weighted image displaying the preoperative appearance of the tumor, while the third and fourth images display the immediate postoperative T1-weighted (contrast-enhanced coronal and pre-contrast sagittal) images, with the light blue arrow showing the lateral approach to the tumor. The patient tolerated the surgery well with no new postoperative language deficit.

*Case 2*: A 40-year-old female patient presented with a nonenhan-cing T2-hyperintense low-grade infiltrating glioma involving the cortex and subcortical white matter of the superomedial aspect of the right precentral gyrus (Fig. 5). Although no histopathology is yet available, the presence of the T2/FLAIR mismatch sign (i.e., high signal diffusely on FSE T2-weighted image but heteroge-neously lower signal on corresponding T2 FLAIR image, as shown on the bottom row images) suggests an IDH1-mutant

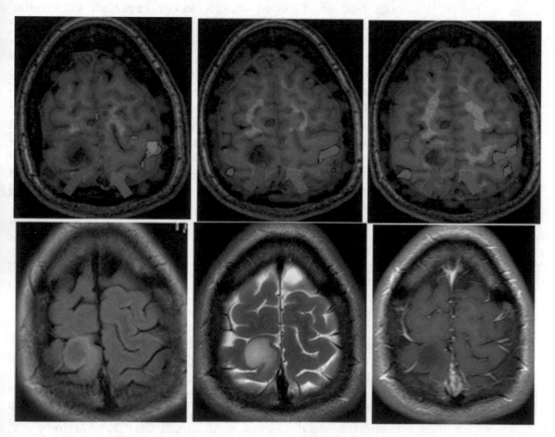

**Fig. 5** Composite language and motor (see main text for the color legend of each task) activation map with BH CVR (low opacity blue) map overlaid (top row). The asymmetry of the foot motor activation map (purple) and BH CVR maps indicated NVU for the right foot motor activation area. In the bottom row, FLAIR, T2, and T1 postcontrast images of this encompassing the superomedial aspect of the right precentral gyrus are shown, suggesting the presence of low-grade IDH1-mutant glioma due to the FLAIR/T2 mismatch sign

low-grade glioma. The postcontrast T1-weighted image is shown on the bottom right.

The composite motor plus language activation map (shown on the top row on three anatomically consecutive images from superior to inferior) displays six different tasks including four language tasks and two motor tasks, each individually color-coded and thresholded (at *T*-values varying from approximately 2.8–3.7): light blue/cyan indicates the rhyming task, magenta indicates a noun–verb semantic association task, red refers to a silent word generation task, and yellow denotes a sentence completion task. Additionally, green depicts hand motor activation during performance of a bilateral simultaneous sequential finger tapping task, while purple depicts activation during performance of a bilateral ankle flexion/extension foot motor task. Superimposed on this composite language activation map is a breath-hold cerebrovascular reactivity (BH CVR) map, which has been thresholded at 0.25 BOLD% signal change and overlaid using 30% opacity in dark blue. Please note that all displayed images follow radiologic convention, whereby the anatomic right cerebral hemisphere is shown in the left lateral aspect of the image and vice versa.

The yellow arrows on all three images on the top row display expected foot motor cortical activation along the superomedial-most aspect of the left precentral gyrus, in purple, but the orange arrows show absence of expected symmetric robust contralateral motor cortical activation. The absent foot motor cortical activation on the right corresponds to a region of abnormally decreased regional BH CVR, reflecting neurovascular uncoupling and associated false-negative activation. The patient had intact left foot and ankle motor function and was thus able to perform the foot motor task with only very mild left foot weakness that was only barely noticeable to the patient.

## 6    Resting-State fMRI for CVR Mapping

Although, in most clinical settings, including in brain tumor presurgical mapping, either a breath-hold technique or exogenous gas administration is typically performed for assessment of cerebrovascular reactivity (CVR), particularly for evaluation of risk of neurovascular uncoupling (NVU), resting-state techniques have also been recently explored for these purposes.

The main advantages of resting-state BOLD methods of cerebrovascular reactivity assessment include the ease of performance, absence of need for patient compliance with breath-hold task instructions, and absence of need for additional equipment and personnel. For example, with breath-hold CVR techniques, regardless of end-expiratory or end-inspiratory approaches and exact details of the chosen paradigm design, the timing of breath-hold and normal breathing blocks and associated patient cooperation and proper understanding of instructions are very important for ensuring that an adequate hypercapnia challenge can be achieved and that general linear model (GLM)-based BOLD data analysis will yield reliable results. Thus, monitoring of patient performance using a respiratory belt and real-time assessment of respiratory bellows is essential for ensuring diagnostic image acquisition. Exogenous gas delivery methods are frequently less patient compliance-dependent but involve equipment such as face masks that are often uncomfortable for patients, and patients have less control over the paradigm administration. There are also logistical problems involved including the need for institutional review board/ethics board approval for "off-label" use, need for specialized equipment for gas delivery and respiratory therapy, or anesthesiology support for patient monitoring. Given these limitations, a passive resting-state BOLD imaging method for CVR assessment has been considered as an attractive potential alternative approach. Although resting-state fMRI is relatively "task-free," it does still require a high degree of cooperation from patients to avoid even minimal head motion and willingness to follow instructions to either keep "eyes open with visual fixation" or "eyes closed without sleeping" for the duration of the scan.

Over the past decade or so, there have been several different approaches that have been proposed as resting-state fMRI-based methods for assessment of CVR. One of the first was the method known as resting-state physiological fluctuation amplitude (RSFA), which was described by Kannurpatti and Biswal in 2008 [31]. In this early study, RSFA values correlated highly with BOLD responses resulting from exogenous 5% $CO_2$ inhalation, as well as with the BOLD responses resulting from breath-holds (BH). The authors suggested that the RSFA could serve as a scaling or calibration factor for motor-evoked BOLD task fMRI activation [31]. In a later study, Kannurpatti SS et al. suggested that RSFA may be not just as effective as BH techniques in assessment of CVR but rather may even surpass the accuracy of BH techniques, particularly in older subjects, where task and physiology-related changes may compromise the accuracy of BH techniques, unlike in younger subjects [32].

However, there has been some controversy regarding the relative accuracy of resting-state-based methods of CVR assessment compared to BH CVR approaches. For example, a 2015 study by I. Lipp et al. comparing repeatability of CVR measures using BH technique and three different analysis approaches (block design, sine-cosine regressor, and $CO_2$ regressor) with those of a resting-state technique suggested that BH CVR analyzed using a $CO_2$ regressor in the GLM analysis yielded the lowest variance and greatest repeatability among the four approaches [33]. It should be noted that Wise RG et al. were the first to describe low-frequency variations in the BOLD signal due to resting-state arterial carbon dioxide fluctuations [34].

More recently, another group has suggested an alternative approach for more quantitative CVR mapping using rs-fMRI, although it has the limitation of requiring simultaneous measurement of end-tidal pCO2 [35]. These investigators compared their resting-state quantitative CVR method, which they called "rs-qCVR," to both RSFA and other resting-state methods such as amplitude of low-frequency fluctuations (ALFF) and a global signal regression method and overall compared effectiveness of all of these resting-state BOLD approaches to the "gold standard" exogenous $CO_2$ administration approach that involved specific end-tidal pCO2 targeting through controlled gas delivery in a breathing circuit. The authors suggested that their approach was more effective than the semiquantitative purely resting-state methods such as RSFA or ALFF in reproducing the CVR measures that were derived through exogenous gas delivery.

However, from a practical standpoint, for presurgical mapping, which remains the only truly accepted clinical application of fMRI in the United States, the main reason for evaluation of CVR is the need to detect the presence of NVU. Such NVU results from local hemodynamic alterations produced by focal brain lesions such as tumors or vascular malformations, which are often associated with aberrant vasculature that is characterized by reduced vascular reactivity and (in high-grade gliomas) increased permeability as well as abnormal flow (in AVMs) and possible steal phenomenon. In these cases of unilateral resectable brain pathology, there is no need for absolute quantitation of CVR, but only a need for assessment of relative CVR. In these situations, detection of abnormal CVR in the vicinity of the brain lesion relative to the CVR in the contralateral cerebral hemisphere is sufficient to determine high risk of NVU. In such instances, there is no need to actually measure end-tidal or arterial pCO2 levels, but simple assessment of the ratio of ipsilesional to contralesional BH CVR is sufficient.

Recent studies have shown that in lieu of BH CVR, resting-state metrics may serve as potential surrogate markers of such NVU (Fig. 6). For example, Agarwal et al. from our group have shown that frequency domain resting-state metrics amplitude of low-frequency fluctuations (ALFF) and fractional ALFF (fALFF) may serve that purpose [36]. ALFF is defined as the total power within a band of the frequency spectrum (typically in the 0.01–0.08 Hz range), and fALFF refers to the ratio of the power spectrum of the selected frequency band to that of the entire frequency range; these metrics were first described by Zang in 2007 and Zou in 2008, respectively [37, 38]. By demonstrating regional decreases in ALFF and fALFF within and immediately around a tumor, compared to normal ipsilesional and contralesional cortical regions, NVU risk can be detected. One important caveat

## Patient 1 -- WHO Grade II Glioma

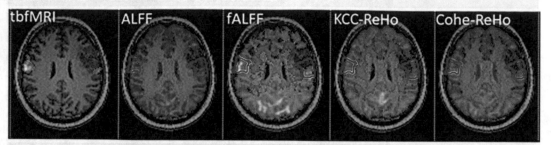

## Patient 2 -- WHO Grade III Anaplastic Astrocytoma

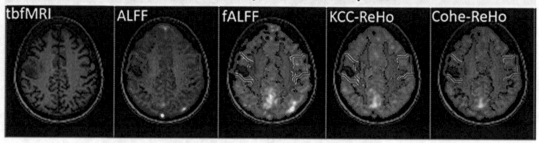

**Fig. 6** Two examples of patients displaying neurovascular uncoupling (NVU) in the sensorimotor network are shown. In each row, the first column shows suprathreshold vertical tongue motor fMRI activation map (tbfMRI) ($Z$-score > 3.5) overlaid on T1-weighted structural images; the second and third column show suprathreshold frequency domain metrics—amplitude of low-frequency fluctuation (ALFF) and fractional ALFF (fALFF)—maps (ALFF > 0.5 and fALFF > 0.4) from resting-state fMRI (rs-fMRI); the fourth and fifth column show suprathreshold regional homogeneity (ReHo)—Kendall's coefficient of concordance (KCC)-ReHo and coherence (Cohe)-ReHo—maps (KCC-ReHo > 0.5 and Cohe-ReHo>0.4) from rs-fMRI. Blue contours display region of interest (ROI) from automatically parcellated combined precentral and postcentral gyri

pertaining to the use of ALFF for NVU detection is that ALFF includes more than simply a CVR component and actually includes a neural component as well. Biswal and colleagues have shown that only the low-frequency portion of the frequency spectrum that is typically interrogated in resting-state fMRI (0.01–0.04 Hz band, but not the higher 0.04–0.08 Hz band) is related to CVR [39]. Nevertheless, ALFF may be a useful marker of NVU. In fact, we have further shown that by utilizing ratios of contralesional mean ALFF to ipsilesional measured ALFF as correction factors, voxel-by-voxel correction of NVU-related BOLD signal loss seen on task fMRI activation maps can be accomplished in many cases of tumor-related NVU (Fig. 7), particularly within the sensorimotor network [40]. This NVU mitigation method holds promise for application in the future to other brain functional networks for which detection may be compromised by lesion-induced NVU.

We have also recently shown [40] that another resting-state measure, known as regional homogeneity (ReHo), may provide similar valuable information regarding NVU potential in brain tumors; specifically, we found that regional decreases in Kendall's

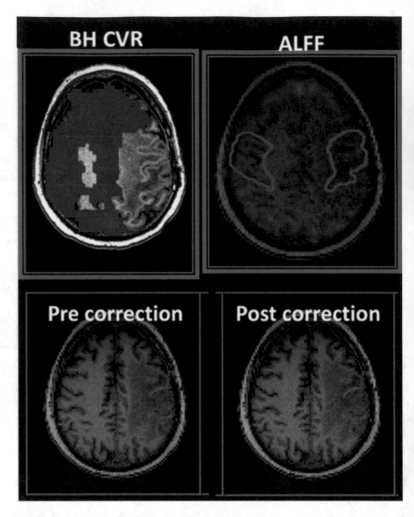

**Fig. 7** A grade IV glioblastoma demonstrating severe sensorimotor network neurovascular uncoupling (NVU) is displayed. A breath-hold cerebrovascular reactivity (BH CVR) map is displayed in the top row on the left, overlaid on T1-weighted anatomic images. The top row, right image, demonstrates the amplitude of low-frequency fluctuation (ALFF) map (Z-score > 0) wherein sensorimotor cortex is contoured with blue-colored ROIs in both contralesional (CL) and ipsilesional (IL) hemispheres. Suprathreshold voxels (Z-score > 2.5) in the primary motor cortex during simultaneous sequential finger tapping (SSFT) task are shown in the bottom row—pre-correction motor activation map on the left and post-ALFF correction map on the right. Notice the newly detectable activation in NVU-affected IL primary sensorimotor cortex on the post-correction map

coefficient of concordance (KCC-ReHo) and coherence regional homogeneity (Cohe-ReHo) corresponding to tumor regions and their immediate vicinity correlated highly with regions of NVU-related decreases in BOLD task activation, suggesting that these metrics may also serve as surrogate markers of NVU [40]. As explained in this paper and originally by Zang and colleagues in

2004, ReHo refers to a measure of how well synchronized the resting-state BOLD time courses within a particular voxel are to those of its immediately neighboring voxels [40, 41].

Resting-state approaches to CVR assessment and NVU detection represent emerging new techniques that appear to be very promising given inherent limitations of more established exogenous gas delivery and breath-hold techniques. More work needs to be done to validate these methods given the relatively small-scale implementation of these approaches to date.

# 7 Future Directions

BH CVR mapping provides a qualitative but clinically feasible approach for assessment of NVU potential in patients with brain tumors and other focal brain lesions who are being evaluated for neurosurgical intervention. Over a decade of cumulative experience in performing BH CVR mapping across multiple academic centers around the world suggests that this technique has led to adoption of BH CVR as a current standard of care in presurgical mapping.

In the near future, new MR sequences such as multiband BOLD that have become routinely available on the new generation of scanners have the potential to make BH CVR more robust. In a recent study by Cohen and Wang, multiband/multi-echo BOLD fMRI has shown increased sensitivity and reproducibility for BH CVR mapping in comparison with the standard single band sequence [21].

The most recently introduced resting-state fMRI approaches will need further validation in larger studies before widespread adoption of resting-state-based CVR mapping for NVU assessment can occur. However, these new approaches are promising, especially for use in patients who are not good candidates for standard BH CVR techniques.

# References

1. Ogawa S, Lee TM, Kay AR et al (1990) Brain magnetic resonance imaging with contrast dependent on blood oxygenation. Proc Natl Acad Sci U S A 87:9868–9872

2. Norris DG (2006) Principles of magnetic resonance assessment of brain function. J Magn Reson Imaging 23:794–807

3. Ogawa S, Lee TM, Nayak AS et al (1990) Oxygenation-sensitive contrast in magnetic resonance image of rodent brain at high magnetic fields. Magn Reson Med 14:68–78

4. Chaudhry AA, Naim S, Gul M et al (2019) Utility of preoperative blood-oxygen-level-dependent functional MR imaging in patients with a central nervous system neoplasm. Radiol Clin N Am 57:1189–1198

5. Vysotski S, Madura C, Swan B et al (2018) Preoperative FMRI associated with decreased mortality and morbidity in brain tumor patients. Interdiscip Neurosurg 13:40–45

6. Black DF, Vachha B, Mian A et al (2017) American society of functional

neuroradiology-recommended fMRI paradigm algorithms for presurgical language assessment. AJNR Am J Neuroradiol 38:E65–E73

7. Attwell D, Buchan AM, Charpak S et al (2010) Glial and neuronal control of brain blood flow (review). Nature 468:232–243

8. Pillai JJ, Mikulis DJ (2015) Cerebrovascular reactivity mapping: an evolving standard for clinical functional imaging. AJNR Am J Neuroradiol 36:7–13

9. Holodny AI, Schulder M, Liu WC et al (2000) The effect of brain tumors on BOLD functional MR imaging activation in the adjacent motor cortex: implications for image-guided neurosurgery. AJNR Am J Neuroradiol 21:1415–1422

10. Ulmer JL, Hacein-Bey L, Mathews VP et al (2004) Lesion-induced pseudo-dominance at functional magnetic resonance imaging: implications for preoperative assessments. Neurosurgery 55:569–579

11. Markus H, Cullinane M (2001) Severely impaired cerebrovascular reactivity predicts stroke and TIA risk in patients with carotid artery stenosis and occlusion. Brain 124:457–467

12. van der Zwan A, Hillen B, Tulleken CA et al (1993) A quantitative investigation of the variability of the major cerebral arterial territories. Stroke 24:1951–1959

13. Mandell DM, Han JS, Poublanc J et al (2008) Selective reduction of blood flow to white matter during hypercapnia corresponds with leukoaraiosis. Stroke 39:1993–1998

14. Mark CI, Fisher JA, Pike GB (2001) Improved fMRI calibration: precisely controlled hyperoxic versus hypercapnic stimuli. NeuroImage 54:1102–1111

15. Ogasawara K, Okuguchi T, Sasoh M et al (2003) Qualitative versus quantitative assessment of cerebrovascular reactivity to acetazolamide using iodine-123-N-isopropyl-p-iodoamphetamine SPECT in patients with unilateral major cerebral artery occlusive disease. AJNR Am J Neuroradiol 24:1090–1095

16. Berthezene Y, Nighoghossian N, Meyer R et al (1998) Can cerebrovascular reactivity be assessed by dynamic susceptibility contrast-enhanced MRI? Neuroradiology 40:1–5

17. Zacà D, Hua J, Pillai JJ (2011) Cerebrovascular reactivity mapping for brain tumor presurgical planning. World J Clin Oncol 2:289–298

18. Kastrup A, Kruger G, Neumann-Haefelin T et al (2001) Assessment of cerebrovascular reactivity with functional magnetic resonance imaging: comparison of CO(2) and breath holding. Magn Reson Imaging 19:13–20

19. van Niftrik CH, Piccirelli M, Bozinov O et al (2016) Fine tuning breath-hold-based cerebrovascular reactivity analysis models. Brain Behav 6:e00426

20. Magon S, Basso G, Farace P et al (2009) Reproducibility of BOLD signal change induced by breath holding. NeuroImage 45:702–712

21. Cohen AD, Wang Y (2019) Improving the assessment of breath-holding induced cerebral vascular reactivity using a multiband multi-echo ASL/BOLD sequence. Sci Rep 9:5079

22. Vesely A, Sasano H, Volgyesi G et al (2001) MRI mapping of cerebrovascular reactivity using square wave changes in end-tidal PCO2. Magn Reson Med 45:1011–1013

23. Mikulis DJ, Krolczyk G, Desal H et al (2005) Preoperative and postoperative mapping of cerebrovascular reactivity in moyamoya disease by using blood oxygen level-dependent magnetic resonance imaging. J Neurosurg 103:347–355

24. Pak RW, Hadjiabadi DH, Senarathna J et al (2017) Implications of neurovascular uncoupling in functional magnetic resonance imaging (fMRI) of brain tumors. J Cereb Blood Flow Metab 37:3475–3487

25. Pillai JJ, Zacà D (2012) Comparison of BOLD cerebrovascular reactivity mapping and DSC MR perfusion imaging for prediction of neurovascular uncoupling potential in brain tumors. Technol Cancer Res Treat 11:361–374

26. Holodny AI, Schulder M, Liu WC et al (1999) Decreased BOLD functional MR activation of the motor and sensory cortices adjacent to a glioblastoma multiforme: implications for image-guided neurosurgery. AJNR Am J Neuroradiol 20:609–612

27. Hou BL, Bradbury M, Peck KK et al (2006) Effect of brain tumor neovasculature defined by rCBV on BOLD fMRI activation volume in the primary motor cortex. NeuroImage 32:489–497

28. Zacà D, Jovicich J, Nadar SR et al (2014) Cerebrovascular reactivity mapping in patients with low grade gliomas undergoing presurgical sensorimotor mapping with BOLD fMRI. J Magn Reson Imaging 40:383–390

29. Hakyemez B, Erdogan C, Ercan I et al (2005) High-grade and low-grade gliomas:

differentiation by using perfusion MR imaging. Clin Radiol 60:493–502

30. Li M, Liu Q, Guo R et al (2020) Perinidal angiogenesis is a predictor for neurovascular uncoupling in the periphery of brain arteriovenous malformations: a task-based and resting-state fMRI study. J Magn Reson Imaging. https://doi.org/10.1002/jmri.27469

31. Kannurpatti SS, Biswal BB (2008) Detection and scaling of task-induced fMRI-BOLD response using resting state fluctuations. NeuroImage 40:1567–1574

32. Kannurpatti SS, Motes MA, Biswal BB et al (2014) Assessment of unconstrained cerebrovascular reactivity marker for large age-range FMRI studies. PLoS One 9(2):e88751

33. Lipp I, Murphy K, Caseras X et al (2015) Agreement and repeatability of vascular reactivity estimates based on a breath-hold task and a resting state scan. NeuroImage 113:387–396

34. Wise RG, Ide K, Poulin MJ et al (2004) Resting fluctuations in arterial carbon dioxide induce significant low frequency variations in BOLD signal. NeuroImage 21:1652–1664

35. Golestani AM, Wei LL, Chen JJ (2016) Quantitative mapping of cerebrovascular reactivity using resting-state BOLD fMRI: validation in healthy adults. NeuroImage 138:147–163

36. Agarwal S, Lu H, Pillai JJ (2017) Value of frequency domain resting-state functional magnetic resonance imaging metrics amplitude of low-frequency fluctuation and fractional amplitude of low-frequency fluctuation in the assessment of brain tumor-induced neurovascular uncoupling. Brain Connect 7:382–389

37. Zang YF, He Y, Zhu CZ et al (2007) Altered baseline brain activity in children with ADHD revealed by resting-state functional MRI. Brain and Development 29:83–91

38. Zou QH, Zhu CZ, Yang Y et al (2008) An improved approach to detection of amplitude of low-frequency fluctuation (ALFF) for resting-state fMRI: fractional ALFF. J Neurosci Methods 172:137–141

39. Biswal BB, Kannurpatti SS, Rypma B (2007) Hemodynamic scaling of fMRI-BOLD signal: validation of low-frequency spectral amplitude as a scalability factor. Magn Reson Imaging 25:1358–1369

40. Agarwal S, Sair HI, Pillai JJ (2017) The resting state fMRI regional homogeneity (ReHo) metrics KCC-ReHo & Cohe-ReHo are valid indicators of tumor-related neurovascular uncoupling. Brain Connect 7:228. https://doi.org/10.1089/brain.2016.0482

41. Zang Y, Jiang T, Lu Y et al (2004) Regional homogeneity approach to fMRI data analysis. NeuroImage 22:394–400

# Chapter 10

## Clinical Translation of Cerebrovascular Reactivity Mapping

### Manus J. Donahue

### Abstract

Cerebrovascular reactivity (CVR), defined broadly as the ability of brain parenchyma to adjust cerebral blood flow and volume in response to altered metabolic demand or a vasoactive stimulus, is being measured with increasing frequency and may have a use for portending infarct development or progression in individuals with cerebrovascular disease. Here, the physiological motivation for performing CVR measurements is presented, followed by safety, feasibility, and administrative considerations for performing such measurements in a clinical setting. Finally, clinical applications are presented in patients with cerebrovascular disease, focusing on how CVR measurements are being used to provide a more complete perspective on cerebral physiology and health.

**Keywords** Cerebrovascular reactivity, Cerebrovascular reserve, Cerebral blood flow, Hemodynamics, Stroke, Neuroimaging

## 1 Introduction

Cerebrovascular disease represents one of the top three causes of death and disability in individuals aged 65 years and older in developed nations and is the most common life-threatening neurological event. Ensuing cerebral tissue death that can arise from cerebrovascular disease, be it acute (i.e., ischemic or hemorrhagic overt stroke) or chronic (silent cerebral infarcts, i.e., infarcts present on neuroimaging but without neurological symptoms), occurs fundamentally from a disruption of the blood supply to cerebral tissue. More specifically, the rate of blood supply to cerebral tissue, commonly measured as cerebral blood flow (CBF, mL blood/100 g tissue/min), reduces as the arterial blood oxygen supply is reduced, either as a result of arterial steno-occlusion or anemia. However, CBF can also be maintained at the microvascular level through relaxation of smooth muscle that lines the cerebral arterioles. Such relaxation of smooth muscle increases local cerebral blood volume (CBV) or the fractional amount of blood relative to tissue. Through this mechanism, CBF can be maintained through the so-called autoregulatory

Jean Chen and Jorn Fierstra (eds.), *Cerebrovascular Reactivity: Methodological Advances and Clinical Applications*, Neuromethods, vol. 175, https://doi.org/10.1007/978-1-0716-1763-2_10, © Springer Science+Business Media, LLC, part of Springer Nature 2022

vasodilation, and as such, assays capable of interrogating the extent of autoregulatory vasodilation can provide a more complete perspective on parenchymal health, and even impending ischemic risk, compared to baseline CBF alone.

This information is now becoming increasingly reported from cerebrovascular reactivity (CVR) measurements, where CVR is defined as the ability of brain parenchyma to adjust CBF in response to altered metabolic demand or a vasoactive stimulus. CVR mapping has been elegantly pursued in research settings for more than 30 years and is emerging as a promising clinical tool, especially for investigating ischemic risk in patients with cerebrovascular disease. The purpose of this chapter is to summarize the relevance and feasibility of performing CVR studies *in a clinical setting*, within the confines of often busy radiological units. This chapter will outline relevance of CVR to acute and chronic cerebrovascular disease, as well as the safety, methodological feasibility, and administrative concerns of performing these measurements.

## 2    Physiological Relevance of Cerebrovascular Reactivity

Before considering the range of procedures for evaluating CVR in clinical applications, it is first useful to understand the relevance of this parameter and its relation to cerebral health. Cerebral ischemia will arise due to reductions in oxygen delivery to tissue and can occur as a result of reduced oxygen delivery without arterial steno-occlusion (e.g., anemia or generalized hypoxemic hypoxia) or decreases in cerebral perfusion pressure (CPP) as a result of arterial steno-occlusion. In cases of reduced oxygen delivery for any of the above reasons, vasodilation of pial arterioles will ensue, resulting in a decrease in vascular resistance and an increase in blood flow. In some cases, these tissue-level compensations may be sufficient to ensure the necessary supply of oxygen, but chronic reductions or acute changes in oxygen delivery may cause these vessels to approach their limit for dilation or exhaust cerebrovascular reserve capacity. When this occurs, this autoregulatory method of compensation is no longer sufficient to maintain adequate oxygen delivery to tissue, and the oxygen extraction fraction (OEF, ratio of oxygen consumed to oxygen delivered) [1] will increase.

As CBF can be maintained over a broad autoregulatory range, the CBF alone often provides an incomplete perspective on how near microvasculature is failing to meet energy demands. However, as microvessels vasodilate and CBV increases, the amount by which CBV may increase further reduces. As such, by challenging the brain parenchyma with a vasodilatory stimulus during a functional neuroimaging acquisition, it is possible to regionally map the CVR, which is proportional to this reserve capacity.

The manner in which microvascular vasodilation occurs in response to reductions in CPP or elevated metabolic demand is governed by cells within the neurovascular unit [2, 3], which consists of neurons, astrocytes, endothelial cells, myocytes, pericytes, and extracellular matrix components. Detailed reviews of the neurovascular unit are beyond the scope of this chapter; however, the relevance of the neurovascular unit for regulating CBF has been described in the literature, and these descriptions have been updated to outline the role of cerebral capillaries [4] and astrocytes [5]. Multiple mechanisms relevant to cerebrovascular disease can lead to modulation of CBF and CBV. Hypoxia, or decreased tissue oxygen pressure ($pO_2$), will lead to increases in CBF and CBV through either direct mechanisms in which the reduced $O_2$ is insufficient to maintain smooth muscle contraction or more indirectly by mediating vasodilatory metabolites. Of particular relevance to many CVR mapping protocols is the reversible effect of $CO_2$ and hydrogen ions ($H^+$) on vascular tone. Hypercapnia (e.g., elevated partial pressure of $CO_2$, $pCO_2$) and hypocapnia (e.g., reduced $pCO_2$) elicit vasodilation and vasoconstriction of the cerebral vasculature, respectively. A detailed summary of in vivo and in situ work on the mechanism of $pCO_2$-induced changes on vascular tone has been reported [6]. What is most relevant for the discussion to follow is that reductions in blood $CO_2$ will lead to vasoconstriction and increases in blood $CO_2$ will lead to vasodilation. The remainder of this document will focus on abilities to modulate and measure these effects directly with vascular stimuli and neuroimaging.

# 3    Stimuli for Eliciting Cerebrovascular Responses

To evaluate CVR, neuroimaging experiments are most commonly performed in conjunction with administration of exogenous agents that alter CBF and CBV. Two major classes of stimuli are generally used: intravenous pharmacological agents and respiratory challenges. In this section, we discuss stimuli from both classes, mechanisms of action, and relevance of each for clinical CVR protocols.

*Acetazolamide*: Acetazolamide (ACZ), a reversible carbonic anhydrase inhibitor which catalyzes the conversion of $CO_2$ into bicarbonate [7], is a common pharmacological agent for eliciting vasodilation [8]. Following intravenous ACZ injection, $pCO_2$ increases, and pH decreases, leading to reduced vascular resistance and vasodilation in compliant microvasculature [9]. ACZ is administered as an intravenous injection, with a dosage of approximately 1000 mg being sufficient for maximal dilative effect [9], and imaging is performed 15–20 min after administration [9, 10]. Higher doses of up to 2000 mg have been utilized [11], and doses are applied based on patient size at a ratio of approximately 16 mg/kg. The advantages of ACZ are that it is generally well-tolerated by patients with minimal side effects and respiratory discomforts and most importantly it is easy to administer as no cooperation is required from the subject [9]. Disadvantages are that it is contra-indicated in a subgroup of patients with kidney disorders, hyper-chloremic acidosis, hypokalemia, and hyponatremia; must be considered in the context of drug interactions, specifically amphe-tamines and other carbonic anhydrase inhibitors; and cannot be easily reversed if patient discomfort is a problem.

*Breath-hold*: The most basic method for modulating $pCO_2$ with a respiratory challenge is through breath-holding [12]. Here subjects are generally instructed to exhale followed by breath-holding for 10–30 s, during which time blood $pCO_2$ increases. Breath-holding has been applied successfully to evaluate volume flow rate changes in healthy adults, and its reproducibility has been evaluated [13], to understand mechanisms of neurovascular coupling using multi-modal MRI in healthy adults [14], and it has been suggested that recurrent ischemic events in patients with carotid occlusion may be predicted by changes in middle cerebral artery (MCA) flow velo-cities during breath-holding [15]. The advantages of breath-holds are that they do not require exogenous contrast agents or equip-ment, and advanced analysis procedures are becoming available for modeling the hemodynamic response function in response to breath-hold or more complex breathing tasks [16, 17]. Disadvan-tages of breath-holds are that they must be short given limited abilities of subjects to hold their breath and in turn allow for few measurements or require without multiple repeats, are sensitive to subject compliance, will induce a gradient rather than constant $pCO_2$ increase over the challenge duration, and are themselves a motor task thereby eliciting neuronal and vascular activity in motor networks.

*Hypercapnic hyperoxia*: A hypercapnic (e.g., $CO_2$ fraction 4–7%) hyperoxic (e.g., $O_2$ fraction >21% and typically 50–95%) stimulus, with balance nitrogen, is also possible and is most common in cancer applications, often with carbogen-5 (e.g., 5% $CO_2$ and 95% $O_2$). Such stimuli have been shown to increase oxygen delivery to tissue due to the vasodilatory effects of hypercapnia and elevated blood oxygen provided by the hyperoxic component; however, these benefits are likely small for typical scans over short stimulus durations [18]. Administration of a **hypercapnic hyperoxic** gas mix-ture for MRI assessment of CVR introduces several experimental confounds because of the **hyperoxic** component including (1) higher venous $HbO_2$ and elevated partial pressure of venous oxygen (**$PvO_2$**) from increased $O_2$ dissolved in plasma, (2) changes in vascular and tissue susceptibility due to increased $HbO_2$, and

(3) decreased blood water $T_1$ due to increased $O_2$ dissolved in plasma in MRI experiments [19]. Advantages of hypercapnic hyperoxia are that this gas mixture can be found readily at most medical centers in the form of carbogen-5, which is FDA-approved, albeit for other indications. Disadvantages of hypercapnic hyperoxia pertain to the more complex effects that the interaction of elevated $pCO_2$ and $pO_2$ may have, including considerations regarding the Haldane effect and changes in oxygen saturation and blood $T_1$ in a manner unrelated to vascular compliance, and these CVR protocols still require research approval and consent at most hospitals, as CVR mapping is considered off-label use of carbogen-5.

## 4  Procedures for Administering Vasoactive Stimuli

While intravenous agents are straightforward to administer, procedures for administering respiratory stimuli have not been standardized across centers. Breath-hold stimuli are generally cued either over the scanner intercom or visually using a projector system and in most cases last 10–30 s depending on subject compliance.

Multiple procedures exist for administering gas challenges; however, a controlled comparison of the most common methods has not been undertaken, especially in patients with cerebrovascular disease. The simplest approach is to obtain pre-mixed hypercapnic normoxic (e.g., 5% $CO_2$/21% $O_2$/74% $N_2$) or hypercapnic hyperoxic (e.g., 5% $CO_2$/95% $O_2$) gases from compressed cylinders provided by suppliers certified to deliver medical-grade mixtures. It should be noted that *clinical-grade* gases are designed for calibration of clinical equipment but not for human consumption, and therefore, *medical-grade* mixtures should be requested, and not all suppliers have certification to produce hypercapnic normoxic mixtures. Gases can be delivered from compressed cylinders located outside the scan room through tubing that enters the scan room or from a Douglas bag inside the scan room. Gases can be delivered through clinical-grade tubing to a clinical-grade oxygen face mask, typically at a flow rate of 10–15 L/min. For calibration of the signal, it is also necessary to simultaneously measure the $EtCO_2$ and in some cases fraction of inspired $O_2$ ($FiO_2$) using a nasal cannula and standard clinical or research monitoring equipment [20, 21]. These setups are relatively easy to implement in clinical settings and elicit $EtCO_2$ changes of approximately 5–7 mmHg for $CO_2$ gas fractions of 5%; however, these changes are dependent on the mask and setup, and inter-subject variation is typically on the order of 1–2 mmHg standard deviations. A simple yet reproducible design has been implemented by Tancredi et al. [20, 22], and this setup comprises an accessible system that can be implemented using clinical-grade equipment and applied in both research and clinical settings at low cost. In addition, the gas can be delivered to the subject using a diving mouthpiece and nose clip [23]. Here the source of the gas is similar, but the subject receives the gas through the mouth with the nose closed. This can in many cases provide a more controlled system, albeit at the cost of some subject discomfort. The major advantages to delivering pre-mixed gas compositions are that the setup is relatively straightforward and pre-scan calibrations or preparations are minimal, and as such, these studies can be performed as part of standard clinical protocols with only a moderate increase in scan time or complexity. The disadvantages primarily pertain to mild discomfort, ensuring that the volunteer can fit in the coil comfortably with the mask or mouthpiece, and some variation in $EtCO_2$ attributable to subject compliance.

More complex gas delivery systems are also possible. Wise et al. [24] have elegantly outlined an end-tidal forcing system, in which gas mixtures are titrated to the subject based on individual physiology and the measured $EtCO_2$ changes. The advantage of this system is that the gas delivery can be assured to be the same in every subject as long as the monitoring equipment is accurate. The disadvantage is that the setup can be cumbersome and is generally not possible to implement for routine clinical scanning. Additionally, computer-controlled rebreathing devices have been developed and exploit similar principles of the subject's specific physiology and respiration to target changes in blood gases to a higher level of accuracy (approximately reported as 1 mmHg accuracy in $CO_2$ targeting for most subjects). Detailed reviews of these devices are available [25]. The disadvantages of these systems are that they require more setup than the standard compressed cylinders and face mask options, require equipment that is currently not FDA-approved, and are more expensive than the above setups. The advantages are that they can provide a more controlled stimulus, which could be especially important when evaluating subtle differences in CVR.

## 5 Safety Data for CVR mapping

Pharmacological induction of vascular changes using ACZ is commonly performed in most major clinical centers, and the vascular response is assessed using SPECT perfusion imaging [26–28]. ACZ has a biological half-life of 2–4 h and is clinically implicated in many patients for treatment of glaucoma, heart failure, epilepsy, and high-altitude sickness. Contraindications are liver or kidney disease, adrenal insufficiency, hyponatremia, hypokalemia, and hyperchloremic acidosis. Drug interactions must also be considered. While sulfonamide allergy has been considered a contraindication for ACZ administration [9], recent literature [29] has shown that

ACZ may be safely utilized in patients with a history of allergy to sulfonamide.

Additionally, CVR measurements using respiratory stimuli and hypercapnia have generally reported minimal or no serious complications, with most complaints related to claustrophobia and general discomfort with the face mask apparatus. In a study of 294 patients (age range = 9–88 years) with cerebrovascular disease, primarily comprised of patients with atherosclerosis, moyamoya, arteriovenous malformation, vasculitis, aneurysm, or dissection, scanned using a hypercapnic normoxic stimulus delivered with a computer-controlled rebreathing apparatus, transient symptoms consisting of shortness of breath, headache, or dizziness were observed in 11.1% of subjects during the hypercapnic stimulus, and no evidence of other neurological or cardiovascular symptoms during or immediately after the task was observed [30]. In a separate study of 92 patients with symptomatic intracranial stenosis scanned using a hypercapnic hyperoxic stimulus delivered from compressed cylinders and clinical-grade oxygen face masks, no immediate stroke-related complications were reported, and longer-term neurological events fell within the range for expected events in this patient population [31]. Major contraindications for hypercapnic respiratory stimuli are chronic obstructive pulmonary disease, inflammatory airway diseases such as asthma, high blood pressure at the time of the exam (generally >200 mmHg systolic) due to the known effect of hypercapnia to increase blood pressure slightly, and low arterial blood oxygen saturation fractions ($\leq$88%).

The emerging data on safety of CVR experiments suggests that these scans pose low risk to chronic ischemic patients; however, larger-scale studies that involve multiple centers, and also with surveillance, will be necessary to define the absolute risk.

# 6  Neuroimaging Approaches for Recording Cerebrovascular Responses

Once a stimulus is chosen, there are several approaches available for measuring the resulting physiological response.

*Positron emission tomography (PET)*: PET is a molecular imaging modality that utilizes a biological compound, typically introduced through intravenous injection or inhalation, that has been tagged with a radioactive isotope to image the distribution of that compound in the body. Of particular interest to CVR imaging, $^{15}O$ is a radioactive isotope of oxygen, and $^{15}O$-water PET can be used to assess CBF [32]. To measure CVR, CBF-weighted images are acquired with $^{15}O$-water at baseline and following administration of the vasoactive stimulus, and CVR is calculated as the relative difference between CBF at baseline and CBF during the stimulus [11]. The advantage of $^{15}O$-water PET is that CBF can be measured directly and quantitatively [11], and PET is generally considered the gold standard of CBF imaging in humans. The major disadvantage of $^{15}$O-water PET is the short half-life of $^{15}$O (approximately 2 min), which limits studies with $^{15}$O-water to specialized centers where on-site cyclotrons are capable of producing this tracer [33]. Additional disadvantages pertain to the requirement for an arterial line for accurate quantification, requirement for ionizing radiation and exogenous contrast, need for an additional MR or CT scan for anatomical localization, and high costs (generally more than threefold higher than MRI, including the radiochemistry costs).

***Single-photon emission computed tomography (SPECT)***: SPECT is a molecular imaging modality that utilizes a biological compound introduced through intravenous injection, inhalation, or ingestion. The most common SPECT compounds utilized for CVR imaging are $^{99m}$Tc-exametazime, $^{99m}$Tc-bicisate, and $^{123}$I-iodoamphetamine [11]. SPECT data using one of these compounds are acquired at baseline and after administration of a vasoactive stimulus, and, similar to PET, CVR is computed as the percent difference between CBF during stimulus and at baseline [34]. The advantages of SPECT are that the radioactive isotopes have much longer half-lives than those utilized in PET. ACZ SPECT is also the most common method for assessing CVR in the clinic, and SPECT contraindications are extremely limited. Disadvantages of SPECT are that it may be performed over 2 days due to the kinetics of the radioisotopes utilized and requires ionizing radiation and contrast agents introduced may cause a moderate rise in blood pressure in some patients [35].

***Transcranial Doppler ultrasound (TCD)***: TCD is a noninvasive technique that utilizes ultrasonic waves and the Doppler effect to measure flow velocity (cm/s). Velocity has been shown to be associated with both CBF and CBV and is computed by measuring the Doppler shift observed in the vessel of interest, such as the MCA [36, 37]. CVR is measured with TCD as the percent change in blood flow velocity after administration of a vasoactive stimulus compared to baseline flow velocity, and CVR measurements made with TCD have been utilized as a marker of stroke risk in asymptomatic carotid artery stenosis [38]. The advantages of TCD are that measurements can be made quickly and with less expensive equipment. The disadvantages are that it provides a surrogate marker of CBF and that the measurements are highly operator dependent with limited spatial coverage [37].

***Arterial spin labeling (ASL) MRI***: ASL MRI is a noninvasive tracer-based technique for quantitative CBF mapping. In ASL, arterial blood water is used as an endogenous, diffusible tracer and is most frequently labeled using either a single radiofrequency pulse (pulsed ASL, PASL) or a train of short radiofrequency pulses (pseudocontinuous ASL, pCASL) [39]. ASL images can provide quantified CBF maps (mL/100 g/min) upon application of the flow-modified Bloch equation [39]. Therefore, dynamic ASL scans where CBF images are acquired on and off vasoactive stimulus can be used to measure CBF for baseline and hyperemic conditions. The advantages of ASL are that it is noninvasive and provides a direct, quantitative measure of CVR by measuring changes in CBF before and during the vasoactive stimulus with an effective temporal resolution of 4–8 s in most scans. The major disadvantage of ASL is that it is limited by the lifetime of the endogenous tracer, which is determined by the longitudinal relaxation time of arterial blood water (~1.2 s at 1.5 T, ~1.6 s at 3.0 T, and ~2.3 s at 7.0 T) [40, 41]. This is detrimental in patients with steno-occlusive arterial disease where blood arrival times are much longer than the tracer lifetimes [42].

*Blood oxygenation level-dependent (BOLD) MRI:* BOLD MRI is the most popular method for functional neuroimaging in humans (Fig. 1). However, unlike ASL, BOLD provides a non-quantitative marker of CVR. BOLD exploits the physiological phenomenon that increases in CBF exceed increases in $CMRO_2$ for strong neuronal or vascular stimulation. Therefore, a larger fraction of oxyhemoglobin ($HbO_2$), relative to deoxyhemoglobin (dHb), is present in capillaries and veins. As $HbO_2$ is diamagnetic and dHb is paramagnetic, this leads to an increase in blood water $T_2$ and $T_2^*$ in capillaries and veins and thus an increase in the surrounding water signal. Therefore, the BOLD signal is spatially localized to draining veins and thus provides only a surrogate marker of CBF and CVR, which co-localize physiologically with the capillary and smooth muscle-lined arteriolar compartments. The BOLD effect is also complex as it reflects a balance of changes in multiple physiological parameters, including CBF, CBV, and $CMRO_2$, and therefore contrast must be interpreted in the context of this hemodynamic window with the understanding that different combinations of parameter changes may yield similar ensemble BOLD responses [43]. The advantages of BOLD are that it is noninvasive and simple to perform and provides a robust signal, and an example of BOLD CVR mapping in cerebrovascular disease is shown in Fig. 1. The primary disadvantage is that it is not qualitative and contains many signal sources with few direct observables.

## 7 Administrative Concerns for Clinical CVR Mapping

Administrative concerns for CVR mapping derive primarily from the appropriate clinical indication for the scan, the time required, and the issues regarding reimbursement. In this chapter, the focus is cerebrovascular disease as, in this population, regional CVR

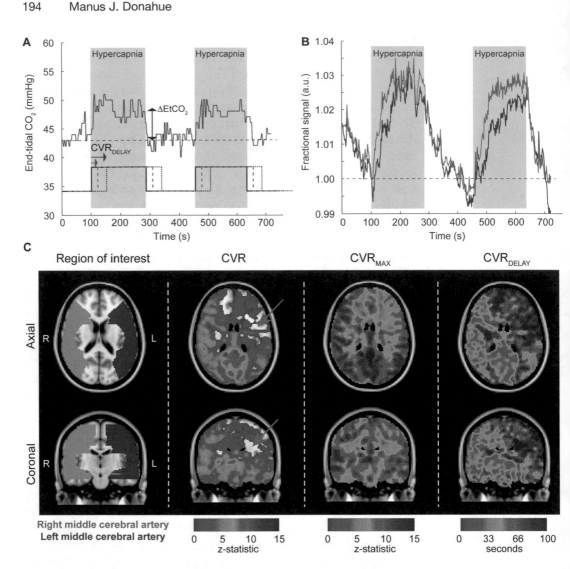

**Fig. 1** Clinical cerebrovascular reactivity (CVR) dynamic mapping using hypercapnic blood oxygenation level-dependent (BOLD) MRI. The stimulus utilizes two 180 s blocks of hypercapnia interleaved with 180 s of normocapnia delivered from FDA-approved equipment. (**a**) This stimulus elicits a 5–7 mmHg increase in end-tidal $CO_2$ (EtCO$_2$), shown here for a 54-year-old female intracranial stenosis patient with right-sided neurological symptoms. CVR describes the $z$-statistic between the voxel time course and the gas-stimulus regressor (black solid). It is also possible to progress the regressor in time (blue and red) and calculate the corresponding $z$-statistic for each time progression element. In this time delay analysis, the maximum $z$-statistic (CVR$_{MAX}$) and time in seconds the regressor has been progressed (CVR$_{Delay}$) can be recorded. Both CVR$_{MAX}$ and CVR$_{Delay}$ may differ in a manner that depends on autoregulatory reserve and collaterals present. (**b**) The blue and red time courses are from the right and left middle cerebral artery (MCA) territories, respectively. The CVR$_{MAX}$ is reduced and CVR$_{Delay}$ lengthened throughout the left MCA territory. (**c**) Axial and coronal representations of the CVR parameters. Lengthened CVR$_{Delay}$ values (arrows) are observed on the side of symptoms

variations are often largest, and also the topography and extent of these variations can influence treatment decisions (i.e., medical management or surgical revascularization). Applications in other populations are also likely relevant, especially neurodegeneration and traumatic brain injury; however, it is currently less clear how CVR findings can be used to influence personalized treatment decisions in these patients.

A standard non-contrast head MRI protocol, generally consisting of T1-weighted, T2-weighted, T2-weighted FLAIR, DWI, and T2*-weighted MRI, can be performed in approximately 20 min. CVR mapping with, for instance, ASL MRI generally has no separate billing code, and therefore, adding the CVR protocol costs only scan time to the radiology department. As there is currently no additional reimbursement for these scans, the support of the scan addition to typical head MRI protocols will vary by Radiology Department and the time demands on imaging equipment. Given this, it useful to keep the time of the scans in mind. For cerebrovascular disease patients where blood arrival time and reactivity times may be longer than in healthy tissue, stimulus periods of 2–3 min are frequently applied, often repeated once, with interval baseline periods. The entire protocol is therefore 10–15 min generally. Often, the actual scan time required is less than the setup time, which will depend on the stimulus of choice. Breath-hold challenges require no setup time outside of instructing the subject, and ACZ and respiratory stimuli administered from compressed cylinders require relatively little setup time, generally less than 5 min which is required for placing the IV or applying a face mask. Computer-controlled rebreathing devices or end-tidal forcing requires more elaborate setups and calibration steps, which may add an additional 10–30 min, depending on the setup used and experience of the user.

The most common clinical CVR mapping methods utilize ACZ administration or a respiratory stimulus administered from compressed cylinders owing to time efficiency; however, specialized centers also use computer-controlled rebreathing devices albeit with additional setup time.

## 8    Applications for Cerebrovascular Reactivity Imaging in Cerebrovascular Disease

CVR mapping has been applied in multiple applications of cerebrovascular disease. This section is intended to highlight clinical unmet needs in the setting of cerebrovascular disease and recent work in which CVR mapping is being applied to address these needs.

*Atherosclerotic arterial steno-occlusive disease*: Reducing stroke-related morbidity requires an improved understanding of early biomarkers that predict stroke and can be used for prescient identification of patients requiring aggressive, preventative therapy [44–46]. For instance, extracranial atherosclerotic disease (ECAD) is a well-known risk factor for stroke; following the North American Stenting vs. Carotid Endarterectomy Trial (NASCET), patients with symptomatic ECAD of the ICA have generally been revascularized surgically, resulting in a low 2–4% 5-year rate of disabling stroke [47, 48]. However, patients with intracranial (IC) atherosclerotic disease (ICAD) comprise approximately 7–24% of new strokes [49–51], and unlike patients with ECAD, appropriate ICAD treatment is less clear as a result of the previously halted SAMMPRIS and VISSIT trials [52, 53]. In these prospective trials, 14–24.1% of ICAD patients treated with stenting and aggressive medical management (AMM), consisting of antiplatelet and statin therapy, experienced a stroke within 30 days, compared to just 5.8–9.4% of patients receiving AMM alone. However, final results from both trials revealed that even in the AMM arms, approximately 12–15% of patients experienced recurrent stroke in 1 year [53]. Due to this high rate of recurrent stroke on standard-of-care therapies, there is an ongoing need to identify sources of hemodynamic impairment in these patients and use this information to personalize therapies.

A prospective study utilizing TCD with hypercapnic stimuli demonstrated that 32% of participants with impaired CVR suffered a stroke or TIA compared to 8% of participants with intact CVR [54]. CVR mapping may also be helpful to assess whether aggressive, albeit controversial, surgical interventions are warranted [55–57]. Utilizing TCD and SPECT with a breath-hold stimulus, it has been shown that superficial temporal artery (STA)-MCA bypass surgery performed in patients with severe steno-occlusive disease and impaired CVR results in a reduction in stroke recurrence [58]. Recent work using hypercapnic BOLD MRI has shown that CVR may be prognostic for recurrent ischemic injury in patients with carotid occlusion [59]. Future work could focus on identifying patients with ICAD at risk of failing standard-of-care AMM and triaging such patients for more aggressive therapies (Fig. 2).

***Non-atherosclerotic arterial steno-occlusive disease***: Moyamoya disease (MMD) is a non-atherosclerotic cerebrovascular condition characterized by progressive stenosis of the supraclinoid ICAs and proximal branches and, frequently, the development of collateral vascular networks [60]. Idiopathic MMD is believed to arise independent of other cerebrovascular conditions and places patients at more than a sevenfold risk increase for stroke [61]. While idiopathic MMD is relatively rare (incidence <1 case/100,000 children in North America), moyamoya syndrome (MMS), which may arise secondary to Down syndrome, sickle cell disease, atherosclerosis, and radiotherapy, shares many phenotypical characteristics as idiopathic MMD and is more common in the general population

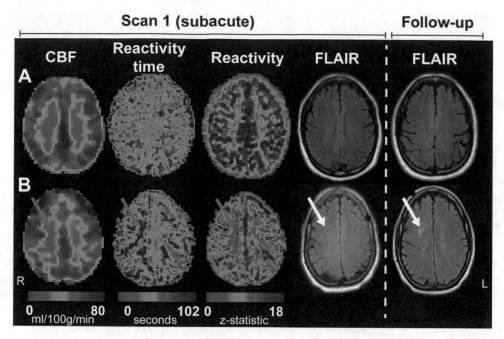

**Fig. 2** Potential of personalized risk stratification using cerebrovascular reactivity mapping in patients with intracranial atherosclerotic disease. (**a**) A 52-year-old female with bilateral M2 stenoses and significant (>70%) left P2 stenosis. CBF, CVR$_{DELAY}$, and CVR are within normal range and symmetric; 1.3-year follow-up shows no new infarcts. (**b**) A 49-year-old female with severe stenosis of the intracranial right ICA. CBF and CVR are reduced, and CVR$_{DELAY}$ is increased, in identical regions, yet there is no infarct present at presentation. At 108 days, new foci of acute infarct within the right centrum semiovale in a watershed distribution are observed (white arrow)

[62, 63]. Such patients have a wide clinical presentation and prognosis, which is hypothesized to depend sensitively on the location and extent of steno-occlusion, parenchymal response via development of collateral vessels, and associated comorbidities [60, 64]. Moyamoya etiology remains largely unknown, animal disease models do not exist, biomarkers that may place patients at highest risk for stroke have not been conclusively established, and randomized trials evaluating preferential utility of surgical revascularization variants have not been performed.

MMS treatment focuses on improving CBF through medical therapy including antiplatelet agents, calcium channel blockers, and more rarely anticoagulants [65, 66]. More aggressive surgical revascularization using either direct bypass or indirect synangiosis procedures is performed in patients with symptoms unresponsive to medical therapy [66] and is now more commonly utilized [67] (Fig. 3). The fundamental clinical question in these patients is whether aggressive surgical procedures are required and, if so, how parenchymal hemodynamic and metabolic responses respond to potentially alter ischemic risk. Addressing these issues likely

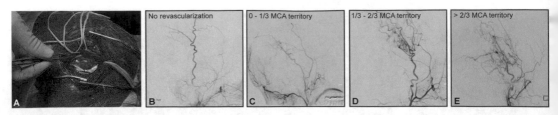

**Fig. 3** (**a**) Example of an indirect encephaloduroarteriosynangiosis (EDAS) revascularization surgery, in which the superficial temporal artery is placed in contact with the brain surface to promote angiogenesis and collateralization. This procedure, however, has varying effectiveness, which on 1-year follow-up on catheter angiography (lateral projections) can lead to poor responses such as (**b**) no revascularization or (**c**) 0–1/3 middle cerebral artery (MCA) territory revascularization or ideally development of new collateral pathways whereby (**d**) 1/3–2/3 or (**e**) >2/3 of the MCA territory is revascularized. CVR mapping is being increasingly utilized to determine collateralization efficacy, as outlined in Fig. 4

requires accurate measurements of how parenchyma compensates for steno-occlusion, including cerebrovascular reserve capacity.

Such non-atherosclerotic conditions represent a popular application of CVR studies, largely due uncertainties with correct management of these patients and abilities of new methods to provide insights regarding extent of impairment and candidacy for revascularization procedures. SPECT imaging originally demonstrated that CBF reactivity was reduced in moyamoya patients [68, 69], and more recently, using MRI in conjunction with hypercapnic stimuli, CVR magnitude has been consistently shown to be reduced, and onset of hemodynamic responses delayed, in patients with moyamoya [70, 71]. CVR timing values have been shown to correlate with arterial circulation times measured from catheter angiography [72], and preliminary scoring systems have been proposed that include CVR as a relevant variable in addition to standard anatomical information from angiography and FLAIR [73]. ASL MRI has also been shown to provide a sensitive marker of hemodynamic impairment in these patients [74], especially when long post-labeling delays are used which account for the delayed blood arrival to tissue [42]. Data using ACZ and SPECT have shown that improvements in cerebral hemodynamics as assessed by CVR following STA-MCA bypass correspond with improvements in patient symptoms [75], and hypercapnic stimuli have been applied to show improvements in CVR following direct and indirect revascularization procedures [70, 76] (Fig. 4).

*Anemia*: A significant mechanism for stroke in patients with sickle cell anemia (SCA, homozygous hemoglobin SS) is hemodynamic imbalance with a reduced supply of oxygen due to reduced oxygen-carrying capacity and HbS presence. Known risk factors for stroke specific to SCA include low baseline hemoglobin, acute chest syndrome, anemic crisis, fever, and hypertension [77, 78]. TCD can be used to assess flow velocity in children with SCA, and elevated MCA flow velocities provide a surrogate for initial stroke risk

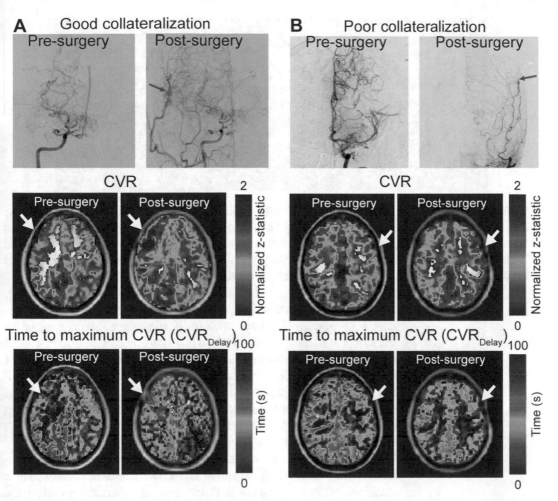

**Fig. 4** Reactivity mapping in patients with moyamoya. (a) A 28-year-old Asian female with moyamoya and successful right-sided EDAS and (b) a 53-year-old Caucasian female with moyamoya and unsuccessful left-sided EDAS. Surgical success was determined based on whether 1/3 or more of the MCA territory was perfused following surgery (e.g., Fig. 3). For the first patient, success of the revascularization procedure is demonstrated by DSA and corresponding increases in CVR and decreases in $CVR_{Delay}$. For the second patient, the revascularization procedure was unsuccessful based on the post-surgery digital subtraction angiography (DSA), and this is indicated by the lack of change seen in the CVR and $CVR_{Delay}$ maps

[79]. However, TCD provides no direct information on CBF and also does not identify increased risk of infarct *recurrence* [79, 80]. To address these limitations, more comprehensive tools to evaluate the spectrum of hemo-metabolic changes that occur in SCA are likely needed. In SCA, when arterial oxygen content decreases, CBF and CBV, and thus oxygen delivery, may increase as a result of autoregulation [81].

Individuals with SCA and without significant vasculopathy typically have chronically increased CBF [81–83] in a manner that depends on several factors including the balance of oxygen-carrying capacity and HbS fraction [84], vasculopathy extent [85], and

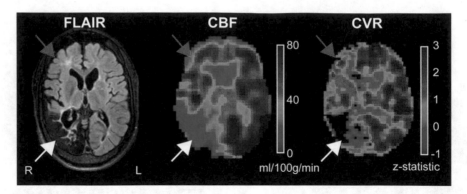

**Fig. 5** Reactivity mapping in a patient with sickle cell disease (SCD). Adult with SCD and right-sided M1 and A1 occlusions. Prior large infarcts are visible in the right MCA and PCA territories (white arrow); here, CBF and CVR are both near zero as expected. The CVR is shown as z-statistic calculated from signal time course and proposed 180 s stimulus paradigm. In right ACA/MCA border zone regions (yellow arrow), CBF is normal to high consistent with SCA physiology and upregulation of CBF to offset anemia; however, CVR is reduced

cerebrovascular reserve [86, 87]. Importantly, these factors influence CBF in opposite ways, with reduced oxygen-carrying capacity leading to increases in CBF, while cerebral vasculopathy and reduced autoregulatory capacity can lower CBF. Thus, CBF is not necessarily specific for the source of impairment, and identical CBF can have distinct underlying hemodynamic mechanisms. Given this, CVR methods hold great promise for additionally evaluating how near patients are to exhausting cerebrovascular reserve with the hypothesis that CBV may be near maximal in many patients (Fig. 5). Indeed, recent work using BOLD MRI in conjunction with a $CO_2$ respiratory stimulus has shown that reduced CVR is associated with reduced cortical thickness in 60 children with sickle cell disease compared to 27 age-matched subjects without sickle cell, most prominently in the left cuneus, the right post-central gyrus, and the right temporal pole [88]. Regional associations also revealed that reduced CVR co-localized with brain regions with high metabolic activity, suggesting that such regions could be more prone to hypoxia-induced damage. Work in a cohort of five patients using MRI showed that CBF decreased and CVR increased following blood transfusion, consistent with increases in oxygen-carrying capacity of blood being associated with lower demands for autoregulation and in turn increases in reserve capacity [89]. ASL MRI before and after ACZ administration can be used to visualize reduced reserve capacity in the face of elevated CBF in individuals with SCA [90].

*Cerebral small vessel disease and aging:* Cerebral small vessel disease (CSVD) is characterized radiologically by the presence of small subcortical lesions, such as white matter hyperintensities and lacunes [91], and vascular endothelial dysfunction has been proposed as a key mechanism for the development of CSVD [92]. The fundamental question in this population is who will develop CSVD and related white matter disease, whether patients with CSVD will progress, and how this impacts cognitive function and symptomatology. A thorough review of the role of CVR imaging in small vessel disease was recently presented by Blair and colleagues [93]. Endothelial dysfunction may result in the stiffening of small blood vessels, causing them to become unreactive to vasoactive stimuli [93]. Thus, CVR imaging may be ideally suited for studying this possible etiology of CSVD. However, currently available data are conflicting [94, 95] on whether CVR is reduced in patients with CSVD, and Blair and colleagues [93] have identified several key factors affecting comparison of CVR studies in this population, including patient heterogeneity, differences in MR acquisition protocols, the vasodilatory stimuli utilized, and the methods for CVR computation. Advancing age is also one of the strongest risk factors for cerebrovascular disease [96]. Whole-brain CBF and regional $CMRO_2$ have been shown to be lower in older adults compared to younger adults [97], and CVR assessed with BOLD MRI and hypercapnic normoxic stimulus has demonstrated age-related decreases in CVR, and this decrease is more prevalent than decreases in CBF [98].

# 9 Conclusions

This chapter summarizes the physiological changes underlying blood flow regulation in response to reduced oxygen delivery to tissue, focusing on the role of cerebral autoregulation. Procedures for measuring surrogates of autoregulation through CVR mapping using PET, SPECT, TCD, and MRI in sequence with pharmacological or respiratory stimuli were summarized, as well as the relevance of these procedures in individuals with ischemic cerebrovascular disease. Translating these methodologies into routine clinical practice at more centers will require consensus on optimized implementation of protocols and analysis procedures, as well as controlled, multicenter trials that outline the sensitivity and specificity of CVR measures on clinical outcomes.

# Acknowledgement

*Funding*: NIH/NINDS 1R01NS07882801A1, NIH/NINDS 1R01NS097763, NIH/NINR 1R01NR01507901, AHA Southeastern affiliate 14GRNT20150004, AHA National affiliate 14CSA20380466

**Conflicts of Interest**: *Manus J. Donahue receives research-related support from Philips Healthcare and grant funding from Pfizer Inc.*

*and is a paid consultant for Pfizer Inc., LymphaTouch, bluebird bio, Novartis, and Global Blood Therapeutics. He is also the CEO of Biosight, LLC, which provides healthcare technology consulting services.*

## References

1. Powers WJ (1991) Cerebral hemodynamics in ischemic cerebrovascular disease. Ann Neurol 29(3):231–240
2. Kim KJ, Filosa JA (2012) Advanced in vitro approach to study neurovascular coupling mechanisms in the brain microcirculation. J Physiol 590(7):1757–1770
3. Hawkins BT, Davis TP (2005) The blood-brain barrier/neurovascular unit in health and disease. Pharmacol Rev 57(2):173–185
4. Itoh Y, Suzuki N (2012) Control of brain capillary blood flow. J Cereb Blood Flow Metab 32 (7):1167–1176
5. Filosa JA, Morrison HW, Iddings JA et al (2016) Beyond neurovascular coupling, role of astrocytes in the regulation of vascular tone. Neuroscience 323:96–109
6. Yoon S, Zuccarello M, Rapoport RM (2012) pCO(2) and pH regulation of cerebral blood flow. Front Physiol 3:365
7. Itada N, Forster RE (1977) Carbonic anhydrase activity in intact red blood cells measured with 18O exchange. J Biol Chem 252 (11):3881–3890
8. Swenson ER (2014) New insights into carbonic anhydrase inhibition, vasodilation, and treatment of hypertensive-related diseases. Curr Hypertens Rep 16(9):467
9. Settakis G, Molnar C, Kerenyi L et al (2003) Acetazolamide as a vasodilatory stimulus in cerebrovascular diseases and in conditions affecting the cerebral vasculature. Eur J Neurol 10(6):609–620
10. Federau C, Christensen S, Zun Z et al (2017) Cerebral blood flow, transit time, and apparent diffusion coefficient in moyamoya disease before and after acetazolamide. Neuroradiology 59(1):5–12
11. Boles Ponto LL, Schultz SK, Leonard Watkins G et al (2004) Technical issues in the determination of cerebrovascular reserve in elderly subjects using 15O-water PET imaging. NeuroImage 21(1):201–210
12. Urback AL, MacIntosh BJ, Goldstein BI (2017) Cerebrovascular reactivity measured by functional magnetic resonance imaging during breath-hold challenge: a systematic review. Neurosci Biobehav Rev 79:27–47
13. de Boorder MJ, Hendrikse J, van der Grond J (2004) Phase-contrast magnetic resonance imaging measurements of cerebral autoregulation with a breath-hold challenge: a feasibility study. Stroke 35(6):1350–1354
14. Donahue MJ, Stevens RD, de Boorder M et al (2009) Hemodynamic changes after visual stimulation and breath holding provide evidence for an uncoupling of cerebral blood flow and volume from oxygen metabolism. J Cereb Blood Flow Metab 29(1):176–185
15. Vernieri F, Pasqualetti P, Passarelli F et al (1999) Outcome of carotid artery occlusion is predicted by cerebrovascular reactivity. Stroke 30(3):593–598
16. Pinto J, Jorge J, Sousa I et al (2016) Fourier modeling of the BOLD response to a breath-hold task: optimization and reproducibility. NeuroImage 135:223–231
17. Bright MG, Bulte DP, Jezzard P et al (2009) Characterization of regional heterogeneity in cerebrovascular reactivity dynamics using novel hypocapnia task and BOLD fMRI. NeuroImage 48(1):166–175
18. Ashkanian M, Borghammer P, Gjedde A et al (2008) Improvement of brain tissue oxygenation by inhalation of carbogen. Neuroscience 156(4):932–938
19. Faraco CC, Strother MK, Siero JC et al (2015) The cumulative influence of hyperoxia and hypercapnia on blood oxygenation and R*(2). J Cereb Blood Flow Metab 35(12):2032–2042
20. Tancredi FB, Lajoie I, Hoge RD (2014) A simple breathing circuit allowing precise control of inspiratory gases for experimental respiratory manipulations. BMC Res Notes 7:235
21. Donahue MJ, Faraco CC, Strother MK et al (2014) Bolus arrival time and cerebral blood flow responses to hypercarbia. J Cereb Blood Flow Metab 34(7):1243–1252
22. Tancredi FB, Lajoie I, Hoge RD (2015) Test-retest reliability of cerebral blood flow and blood oxygenation level-dependent responses to hypercapnia and hyperoxia using dual-echo pseudo-continuous arterial spin labeling and step changes in the fractional composition of inspired gases. J Magn Reson Imaging 42 (4):1144–1157

23. Yezhuvath US, Lewis-Amezcua K, Varghese R et al (2009) On the assessment of cerebrovascular reactivity using hypercapnia BOLD MRI. NMR Biomed 22(7):779–786

24. Wise RG, Pattinson KT, Bulte DP et al (2007) Dynamic forcing of end-tidal carbon dioxide and oxygen applied to functional magnetic resonance imaging. J Cereb Blood Flow Metab 27 (8):1521–1532

25. Fierstra J, Sobczyk O, Battisti-Charbonney A et al (2013) Measuring cerebrovascular reactivity: what stimulus to use? J Physiol 591 (23):5809–5821

26. Bonte FJ, Devous MD, Reisch JS (1988) The effect of acetazolamide on regional cerebral blood flow in normal human subjects as measured by single-photon emission computed tomography. Investig Radiol 23(8):564–568

27. Chollet F, Celsis P, Clanet M et al (1989) SPECT study of cerebral blood flow reactivity after acetazolamide in patients with transient ischemic attacks. Stroke 20(4):458–464

28. Suga Y, Ogasawara K, Saito H et al (2007) Preoperative cerebral hemodynamic impairment and reactive oxygen species produced during carotid endarterectomy correlate with development of postoperative cerebral hyperperfusion. Stroke 38(10):2712–2717

29. Platt D, Griggs RC (2012) Use of acetazolamide in sulfonamide-allergic patients with neurologic channelopathies. Arch Neurol 69 (4):527–529

30. Spano VR, Mandell DM, Poublanc J et al (2013) CO2 blood oxygen level-dependent MR mapping of cerebrovascular reserve in a clinical population: safety, tolerability, and technical feasibility. Radiology 266 (2):592–598

31. Donahue MJ, Dethrage LM, Faraco CC et al (2014) Routine clinical evaluation of cerebrovascular reserve capacity using carbogen in patients with intracranial stenosis. Stroke 45 (8):2335–2341

32. Ter-Pogossian MM, Herscovitch P (1985) Radioactive oxygen-15 in the study of cerebral blood flow, blood volume, and oxygen metabolism. Semin Nucl Med 15(4):377–394

33. Hsu B (2013) PET tracers and techniques for measuring myocardial blood flow in patients with coronary artery disease. J Biomed Res 27 (6):452–459

34. Hashimoto A, Mikami T, Komatsu K et al (2017) Assessment of hemodynamic compromise using computed tomography perfusion in combination with 123I-IMP single-photon emission computed tomography without acetazolamide challenge test. J Stroke Cerebrovasc Dis 26(3):627–635

35. Ma J, Mehrkens JH, Holtmannspoetter M et al (2007) Perfusion MRI before and after acetazolamide administration for assessment of cerebrovascular reserve capacity in patients with symptomatic internal carotid artery (ICA) occlusion: comparison with 99mTc-ECD SPECT. Neuroradiology 49(4):317–326

36. Patrick JT, Fritz JV, Adamo JM et al (1996) Phase-contrast magnetic resonance angiography for the determination of cerebrovascular reserve. J Neuroimaging 6(3):137–143

37. Pindzola RR, Balzer JR, Nemoto EM et al (2001) Cerebrovascular reserve in patients with carotid occlusive disease assessed by stable xenon-enhanced ct cerebral blood flow and transcranial Doppler. Stroke 32(8):1811–1817

38. Silvestrini M, Vernieri F, Pasqualetti P et al (2000) Impaired cerebral vasoreactivity and risk of stroke in patients with asymptomatic carotid artery stenosis. JAMA 283 (16):2122–2127

39. Alsop DC, Detre JA, Golay X et al (2015) Recommended implementation of arterial spin-labeled perfusion MRI for clinical applications: a consensus of the ISMRM perfusion study group and the European consortium for ASL in dementia. Magn Reson Med 73 (1):102–116

40. Lu H, Clingman C, Golay X et al (2004) Determining the longitudinal relaxation time (T1) of blood at 3.0 Tesla. Magn Reson Med 52(3):679–682

41. Rane SD, Gore JC (2013) Measurement of T1 of human arterial and venous blood at 7T. Magn Reson Imaging 31(3):477–479

42. Fan AP, Guo J, Khalighi MM et al (2017) Long-delay arterial spin labeling provides more accurate cerebral blood flow measurements in moyamoya patients: a simultaneous positron emission tomography/MRI study. Stroke 48(9):2441–2449

43. Blicher JU, Stagg CJ, O'Shea J et al (2012) Visualization of altered neurovascular coupling in chronic stroke patients using multimodal functional MRI. J Cereb Blood Flow Metab 32(11):2044–2054

44. Goldstein LB, American Heart Association, American Stroke Association (2009) A primer on stroke prevention and treatment: an overview based on AHA/ASA guidelines. Wiley-Blackwell, Chichester; Hoboken, NJ. xi, 263 p

45. González RG (2005) Acute ischemic stroke: imaging and intervention. Springer, Berlin; New York, NY. xii, 268 p

46. Lin W, An H, Ford AL et al (2013) MR imaging of oxygen extraction and neurovascular coupling. Stroke 44(6 Suppl 1):S61–S64

47. National Institute of Neurological Disorders and Stroke. Stroke and Trauma Division. North American Symptomatic Carotid Endarterectomy Trial (NASCET) investigators (1991) Clinical alert: benefit of carotid endarterectomy for patients with high-grade stenosis of the internal carotid artery. Stroke 22 (6):816–817

48. Reed AB, Gaccione P, Belkin M et al (2003) Preoperative risk factors for carotid endarterectomy: defining the patient at high risk. J Vasc Surg 37(6):1191–1199

49. Kasner SE, Gorelick PB (2004) Prevention and treatment of ischemic stroke. Butterworth-Heinemann, Philadelphia, PA. xv, 416 p

50. Ovbiagele B, Cruz-Flores S, Lynn MJ et al (2008) Early stroke risk after transient ischemic attack among individuals with symptomatic intracranial artery stenosis. Arch Neurol 65 (6):733–737

51. Famakin BM, Chimowitz MI, Lynn MJ et al (2009) Causes and severity of ischemic stroke in patients with symptomatic intracranial arterial stenosis. Stroke 40(6):1999–2003

52. Zaidat OO, Fitzsimmons BF, Woodward BK et al (2015) Effect of a balloon-expandable intracranial stent vs medical therapy on risk of stroke in patients with symptomatic intracranial stenosis: the VISSIT randomized clinical trial. JAMA 313(12):1240–1248

53. Derdeyn CP, Chimowitz MI, Lynn MJ et al (2014) Aggressive medical treatment with or without stenting in high-risk patients with intracranial artery stenosis (SAMMPRIS): the final results of a randomised trial. Lancet 383 (9914):333–341

54. Kleiser B, Widder B (1992) Course of carotid artery occlusions with impaired cerebrovascular reactivity. Stroke 23(2):171–174

55. Attye A, Villien M, Tahon F et al (2014) Normalization of cerebral vasoreactivity using BOLD MRI after intravascular stenting. Hum Brain Mapp 35(4):1320–1324

56. Mandell DM, Han JS, Poublanc J et al (2011) Quantitative measurement of cerebrovascular reactivity by blood oxygen level-dependent MR imaging in patients with intracranial stenosis: preoperative cerebrovascular reactivity predicts the effect of extracranial-intracranial bypass surgery. AJNR Am J Neuroradiol 32 (4):721–727

57. Bouvier J, Detante O, Tahon F et al (2015) Reduced CMRO(2) and cerebrovascular reserve in patients with severe intracranial arterial stenosis: a combined multiparametric qBOLD oxygenation and BOLD fMRI study. Hum Brain Mapp 36(2):695–706

58. Low SW, Teo K, Lwin S et al (2015) Improvement in cerebral hemodynamic parameters and outcomes after superficial temporal artery-middle cerebral artery bypass in patients with severe stenoocclusive disease of the intracranial internal carotid or middle cerebral arteries. J Neurosurg 123(3):662–669

59. Goode SD, Altaf N, Munshi S et al (2016) Impaired cerebrovascular reactivity predicts recurrent symptoms in patients with carotid artery occlusion: a hypercapnia BOLD fMRI study. AJNR Am J Neuroradiol 37(5):904–909

60. Scott RM, Smith ER (2009) Moyamoya disease and moyamoya syndrome. N Engl J Med 360(12):1226–1237

61. Hallemeier CL, Rich KM, Grubb RL Jr et al (2006) Clinical features and outcome in North American adults with moyamoya phenomenon. Stroke 37(6):1490–1496

62. Kassim AA, DeBaun MR (2013) Sickle cell disease, vasculopathy, and therapeutics. Annu Rev Med 64:451–466

63. Phi JH, Wang KC, Lee JY et al (2015) Moyamoya syndrome: a window of moyamoya disease. J Kr Neurosurg Soc 57(6):408–414

64. Achrol AS, Guzman R, Lee M et al (2009) Pathophysiology and genetic factors in moyamoya disease. Neurosurg Focus 26(4):E4

65. Scott RM, Smith JL, Robertson RL et al (2004) Long-term outcome in children with moyamoya syndrome after cranial revascularization by pial synangiosis. J Neurosurg 100 (2 Suppl):142–149

66. Scott RM (2000) Moyamoya syndrome: a surgically treatable cause of stroke in the pediatric patient. Clin Neurosurg 47:378–384

67. Ikezaki K (2000) Rational approach to treatment of moyamoya disease in childhood. J Child Neurol 15(5):350–356

68. Tatemichi TK, Prohovnik I, Mohr JP et al (1988) Reduced hypercapnic vasoreactivity in moyamoya disease. Neurology 38 (10):1575–1581

69. Hoshi H, Ohnishi T, Jinnouchi S et al (1994) Cerebral blood flow study in patients with moyamoya disease evaluated by IMP SPECT. J Nucl Med 35(1):44–50

70. Donahue MJ, Strother MK, Lindsey KP et al (2015) Time delay processing of hypercapnic fMRI allows quantitative parameterization of cerebrovascular reactivity and blood flow delays. J Cereb Blood Flow Metab 36:1767

71. Liu P, Welch BG, Li Y et al (2017) Multiparametric imaging of brain hemodynamics and

function using gas-inhalation MRI. Neuro-Image 146:715–723

72. Donahue MJ, Ayad M, Moore R et al (2013) Relationships between hypercarbic reactivity, cerebral blood flow, and arterial circulation times in patients with moyamoya disease. J Magn Reson Imaging 38(5):1129–1139

73. Ladner TR, Donahue MJ, Arteaga DF et al (2017) Prior Infarcts, Reactivity, and Angiography in Moyamoya Disease (PIRAMD): a scoring system for moyamoya severity based on multimodal hemodynamic imaging. J Neurosurg 126(2):495–503

74. Yun TJ, Paeng JC, Sohn CH et al (2015) Monitoring cerebrovascular reactivity through the use of arterial spin labeling in patients with moyamoya disease. Radiology 278:141865

75. Kawabori M, Kuroda S, Nakayama N et al (2013) Effective surgical revascularization improves cerebral hemodynamics and resolves headache in pediatric Moyamoya disease. World Neurosurg 80(5):612–619

76. Sam K, Poublanc J, Sobczyk O et al (2015) Assessing the effect of unilateral cerebral revascularisation on the vascular reactivity of the non-intervened hemisphere: a retrospective observational study. BMJ Open 5(2):e006014

77. Ohene-Frempong K, Weiner SJ, Sleeper LA et al (1998) Cerebrovascular accidents in sickle cell disease: rates and risk factors. Blood 91 (1):288–294

78. Balkaran B, Char G, Morris JS et al (1992) Stroke in a cohort of patients with homozygous sickle cell disease. J Pediatr 120 (3):360–366

79. Adams RJ (2005) TCD in sickle cell disease: an important and useful test. Pediatr Radiol 35 (3):229–234

80. Adams RJ, McKie VC, Brambilla D et al (1998) Stroke prevention trial in sickle cell anemia. Control Clin Trials 19(1):110–129

81. Vorstrup S, Lass P, Waldemar G et al (1992) Increased cerebral blood flow in anemic patients on long-term hemodialytic treatment. J Cereb Blood Flow Metab 12(5):745–749

82. Prohovnik I, Hurlet-Jensen A, Adams R et al (2009) Hemodynamic etiology of elevated flow velocity and stroke in sickle-cell disease. J Cereb Blood Flow Metab 29(4):803–810

83. Gevers S, Nederveen AJ, Fijnvandraat K et al (2012) Arterial spin labeling measurement of cerebral perfusion in children with sickle cell disease. J Magn Reson Imaging 35 (4):779–787

84. Hurlet-Jensen AM, Prohovnik I, Pavlakis SG et al (1994) Effects of total hemoglobin and hemoglobin S concentration on cerebral blood

flow during transfusion therapy to prevent stroke in sickle cell disease. Stroke 25 (8):1688–1692

85. Arkuszewski M, Krejza J, Chen R et al (2014) Sickle cell anemia: intracranial stenosis and silent cerebral infarcts in children with low risk of stroke. Adv Med Sci 59(1):108–113

86. Derdeyn CP, Videen TO, Yundt KD et al (2002) Variability of cerebral blood volume and oxygen extraction: stages of cerebral haemodynamic impairment revisited. Brain J Neurol 125(Pt 3):595–607

87. Gupta A, Chazen JL, Hartman M et al (2012) Cerebrovascular reserve and stroke risk in patients with carotid stenosis or occlusion: a systematic review and meta-analysis. Stroke 43 (11):2884–2891

88. Kim JA, Leung J, Lerch JP et al (2016) Reduced cerebrovascular reserve is regionally associated with cortical thickness reductions in children with sickle cell disease. Brain Res 1642:263–269

89. Kosinski PD, Croal PL, Leung J et al (2017) The severity of anaemia depletes cerebrovascular dilatory reserve in children with sickle cell disease: a quantitative magnetic resonance imaging study. Br J Haematol 176(2):280–287

90. Vaclavu L, van der Land V, Heijtel DF et al (2016) In vivo T1 of blood measurements in children with sickle cell disease improve cerebral blood flow quantification from arterial spin-labeling MRI. AJNR Am J Neuroradiol 37(9):1727–1732

91. Wardlaw JM, Smith EE, Biessels GJ et al (2013) Neuroimaging standards for research into small vessel disease and its contribution to ageing and neurodegeneration. Lancet Neurol 12(8):822–838

92. Wardlaw JM, Smith C, Dichgans M (2013) Mechanisms of sporadic cerebral small vessel disease: insights from neuroimaging. Lancet Neurol 12(5):483–497

93. Blair GW, Doubal FN, Thrippleton MJ et al (2016) Magnetic resonance imaging for assessment of cerebrovascular reactivity in cerebral small vessel disease: a systematic review. J Cereb Blood Flow Metab 36(5):833–841

94. Conijn MM, Hoogduin JM, van der Graaf Y et al (2012) Microbleeds, lacunar infarcts, white matter lesions and cerebrovascular reactivity -- a 7 T study. NeuroImage 59 (2):950–956

95. Hund-Georgiadis M, Zysset S, Naganawa S et al (2003) Determination of cerebrovascular reactivity by means of FMRI signal changes in cerebral microangiopathy: a correlation with

morphological abnormalities. Cerebrovasc Dis 16(2):158–165

96. Nagata K, Yamazaki T, Takano D et al (2016) Cerebral circulation in aging. Ageing Res Rev 30:49–60

97. De Vis JB, Hendrikse J, Bhogal A et al (2015) Age-related changes in brain hemodynamics; A calibrated MRI study. Hum Brain Mapp 36 (10):3973–3987

98. Lu H, Xu F, Rodrigue KM et al (2011) Alterations in cerebral metabolic rate and blood supply across the adult lifespan. Cereb Cortex 21 (6):1426–1434

# Chapter 11

# Recent Advances and Future Directions: Clinical Applications of Intraoperative BOLD-MRI CVR

Giovanni Muscas, Christiaan Hendrik Bas van Niftrik, Martina Sebök, Alessandro Della Puppa, Luca Regli, and Jorn Fierstra

## Abstract

Blood oxygenation level-dependent (BOLD) magnetic resonance imaging (MRI) sequences have gained widespread interest in recent years as an effective way to investigate cerebrovascular reactivity (CVR, a measure of the hemodynamic state of the brain) with a high spatial and temporal resolution. The clinical relevance of CVR in diverse pathologies has been widely tested, especially ischemic cerebrovascular diseases. Here, its importance has been confirmed both preoperatively for a better stratification risk and postoperatively to evaluate the effectiveness of revascularization procedures. Recently, CVR assessments have shown interesting findings in neuro-oncology. The ability to obtain this information intraoperatively is, however, novel and has not been tested. We report our first experience with this intraoperative technique in vascular and oncologic neurosurgical patients and discuss the results of its feasibility and the possible developments of the intraoperative employment of BOLD-CVR.

Key words Cerebrovascular reactivity, Functional MRI, Intraoperative, STA-MCA, Brain gliomas

## 1 Introduction

Brain vessels are physiologically able to respond to a vasoactive stimulus and modulate the cerebral blood flow (CBF) by changing their caliber as needed. This capacity is called cerebrovascular reactivity (CVR), and its effective functioning is crucial to supply an adequate amount of oxygenated blood to the brain despite wide variations of perfusion pressure (cerebrovascular reserve) [1, 2].

Various neurological and neurosurgical diseases can stress this capacity beyond a maximum that cannot be further compensated, leading to exhaustion of the cerebrovascular reserve, which is measured as an impairment in CVR [3]. Many studies have

---

The original version of this chapter was revised. The correction to this chapter is available at https://doi.org/10.1007/978-1-0716-1763-2_12

Jean Chen and Jorn Fierstra (eds.), *Cerebrovascular Reactivity: Methodological Advances and Clinical Applications*, Neuromethods, vol. 175, https://doi.org/10.1007/978-1-0716-1763-2_11, © Springer Science+Business Media, LLC, part of Springer Nature 2022, Corrected Publication 2022

highlighted the importance of impaired CVR as a measure of the hemodynamic stress the brain is undergoing in a particular anatomical area [4–9], and the clinical relevance of CVR in predicting stroke risk has been broadly confirmed. Currently, CVR can be investigated with different techniques, allowing risk stratification and providing valuable information for therapeutic decision-making [9, 10]. Mainly, CVR is assessed by applying a vasoactive stimulus and measuring the resulting changes at the brain level, either by CBF modifications or by the surrogate of blood flow. As a vasoactive stimulus, drugs or endogenous substances like $CO_2$ are used [2]: variations in the arterial $CO_2$ concentrations can be obtained by induced or voluntary apnea [2, 11]. For measurement of hemodynamic changes, various methods can be used, like Doppler sonography [9], arterial spin labeling (ASL) [12], or blood oxygenation level-dependent cerebrovascular reactivity (BOLD-CVR) [3]. BOLD-CVR can detect CVR changes with high spatial and temporal resolution: due to the deoxyhemoglobin paramagnetic properties, the BOLD signal changes following the increased deoxyhemoglobin washout associated with vasodilation and subsequent higher blood flows [2, 13].

In neurosurgery, BOLD-CVR assessment has found many applications both pre- and postoperatively [5, 6, 14–18], allowing to gain relevant information for prognosis and postoperative evaluation. Intraoperative evaluation of BOLD signal changes associated with CVR variations, however, has never been attempted, despite its potential role in giving new information on hemodynamic changes associated with surgical maneuvers.

We have recently proposed and tested an intraoperative CVR assessment to obtain new information, which could potentially help during the surgical procedure [19, 20]. Here, we describe the rationale, techniques, relevance of knowledge gained, limitations, and potential development of intraoperative BOLD-CVR assessment.

## 2   Materials and Methods

The cantonal ethics board of the Canton of Zurich, Switzerland (KEK-ZH-Nr. 2012-0427), approved patients' recruitment, and all participants signed informed consent. Subjects undergoing intraoperative MR scans during surgery for vascular or oncological diseases were recruited. In the operating setting, sedated and intubated patients were transferred from the operating theater to the adjacent MR suite, following a previously described protocol used at our institution [20, 21], and scanned on a 3-T device, obtaining BOLD and T1-weighted without contrast enhancement plus other sequences according to the specific patient's case (i.e., T1-weighted with contrast, T2-weighted, fluid attenuation inversion recovery [FLAIR] and susceptibility-weighted imaging [SWI] sequences for tumor patients, TOF MRA or diffusion-weighted imaging [DWI] in vascular diseases). The technical details about acquisition parameters can be found in our previous publications [16, 19, 20]. Hypercapnia

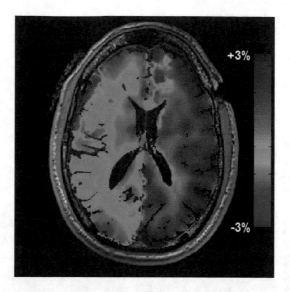

**Fig. 1** Example of intraoperative BOLD-CVR color-coded map after bypass flow augmentation surgery. Intraoperative BOLD-CVR assessment after STA-MCA bypass for flow augmentation in a patient with occlusion of the right internal carotid artery. Areas with persisting severely impaired hemodynamic status are depicted in blue. The color scale refers to the mean percentage change in BOLD signal per voxel between baseline and hypercapnic state (red indicating normal CVR toward blue indicating severely impaired CVR)

was used as a vasoactive stimulus for CVR assessment: considering the technical issues associated with obtaining controlled iso-oxic hypercapnic stimuli, which are usually evoked in awake patients (further details are described elsewhere [2]), breath-holding tasks were performed to obtain $CO_2$ raises in sedated and unconscious patients. During the acquisition of the BOLD sequences, hypercapnia was induced by three 44-s apnea cycles obtained by halting the ventilator, followed by hyperventilation to return to the subject's baseline $CO_2$. Between each hypercapnic challenge, an 88-s baseline $CO_2$ period is maintained. This allowed obtaining satisfactory $CO_2$ oscillations and, therefore, adequate vasoactive stimuli resulting in coherent BOLD signal fluctuations [20].

Pre-processing and data analysis were performed following a method described previously [20, 22, 23]. CVR could be thus calculated on a voxel-by-voxel basis and measured as the mean %BOLD signal change between baseline and hypercapnia. For optimal depiction, the % change values were color-coded and overlaid on the T1-weighted anatomical scan acquired during the same session (*see* Fig. 1).

# 3    Results

The patients included are summarized in Table 1 (see also Figs. 2 and 3). As a first relevant finding, none of the patients displayed

**Table 1**
Patient's sex, age, diagnosis, and intraoperative CVR values of all the patients undergoing intraoperative BOLD-CVR assessment included in previous studies. CVR values are calculated as the mean %BOLD signal change between baseline and the hypercapnic (breath–hold) challenge

| | Sex/age | Diagnosis | CVR values of the whole brain | Affected hemisphere | Unaffected hemisphere | Difference between the affected and unaffected hemisphere (%) |
|---|---|---|---|---|---|---|
| 1 | M 43 | Anaplastic astrocytoma (grade III WHO) | 0.9 | 1.18 | 0.94 | 125.5 |
| 2 | M 31 | Oligoastrocytoma (grade III WHO) | 0.13 | 0.83 | 0.18 | 461.1 |
| 3 | M 18 | Inconclusive | 1.12 | 1.24 | 1.05 | 118.1 |
| 4 | M 51 | Glioblastoma (grade IV WHO) | 1.01 | 1.02 | 0.8 | 127.5 |
| 5 | F 41 | Anaplastic astrocytoma (grade III WHO) | 0.95 | 1.01 | 0.89 | 113.5 |
| 6 | M 38 | Anaplastic oligoastrocytoma (grade III WHO) | 0.85 | 0.1 | 0.69 | 14.5 |
| 7 | M 18 | Medulloblastoma | 0.14 | 1.37 | 0.19 | 721.1 |
| 8 | M 46 | Binswanger's disease | 1.35 | 0.48 | 1.33 | 36.1 |
| 9 | M 76 | 3-vessel atherosclerosis | 0.43 | 0.42 | 0.38 | 110.5 |
| 10 | F 42 | Intravasal and mural myxomas | 0.4 | 0.72 | 0.37 | 194.6 |
| 11 | F 68 | ICA occlusion | 1.4 | 1 | 1.95 | 51.3 |
| 12 | F 66 | ICA occlusion | 2.08 | 1.54 | 2.61 | 59.0 |
| 13 | F 49 | Moyamoya disease | 1.44 | 1.64 | 1.22 | 134.4 |
| 14 | F 62 | ICA, MCA, and ACA occlusion | 0.16 | 0.03 | 0.28 | 10.7 |
| 15 | M 50 | ICA occlusion | 0.76 | 0.71 | 0.81 | 87.7 |
| 16 | F 46 | Anaplastic astrocytoma (grade III WHO) | 1.77 | 1.62 | 1.83 | 88.5 |

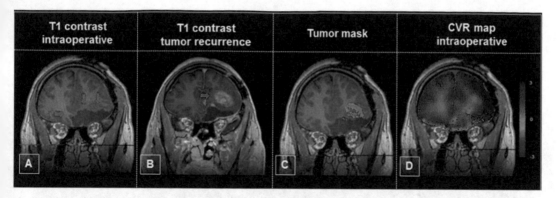

| T1 contrast intraoperative | T1 contrast tumor recurrence | Tumor mask | CVR map intraoperative |

**Fig. 2** Association between intraoperative BOLD-CVR findings and areas of tumor recurrence on follow-up MR imaging. (**a**) Coronal view of an intraoperative contrast-enhanced T1-weighted MR scan depicting the surgical cavity (marked with the green arrows) with no clear contrast enhancement visible; (**b**) follow-up imaging showing tumor recurrence (yellow arrow indicating the contrast-enhancing part, i.e., tumor recurrence); (**c**) the contrast-enhancing area was then automatically masked and overlaid on the intraoperative BOLD-CVR scan done previously (**d**) which, in the area of tumor recurrence, shows lower CVR values compared to the contralateral hemisphere. (Picture adopted from Fierstra et al. [19] Magn Reson Imaging 2016;34:803–8)

intra- or postoperative complications due to the prolonged time for anesthesia or the breath-hold task. Analyzing CVR data showed impaired values for oncologic patients within the lesion. In both hemispheres, an impressive result that was discussed in previous work from our group [16] was observed, and more interestingly, these patients showed lower CVR values in areas associated with tumor recurrence on follow-up [19] (Fig. 2).

In patients undergoing STA-MCA flow augmentation bypass surgery, intraoperative CVR values were, on average, higher after revascularization [24] (Fig. 3).

# 4 Notes

Developing and improving an intraoperative BOLD-CVR assessment aims at offering a new tool to aid surgery and influence intraoperative strategy and decision-making by providing early feedback on the hemodynamic state of the brain. CVR is a measure of the vessels' ability to modulate blood flow (CBF) [1]. However, variations in the BOLD signal are dependent on other physiological parameters than the singular CBF, like cerebral blood volume (CBV), hematocrit, and cerebral metabolic rate of oxygen ($CMRO_2$) consumption [13, 25], and this should be considered when evaluating intraoperative changes in CVR.

However, it has been widely confirmed that cerebrovascular reserve can be exhausted in some areas undergoing relevant hemodynamic stress. In these areas, a vasoactive stimulus produces a "steal" blood flow to regions with consumed reserve capacity to

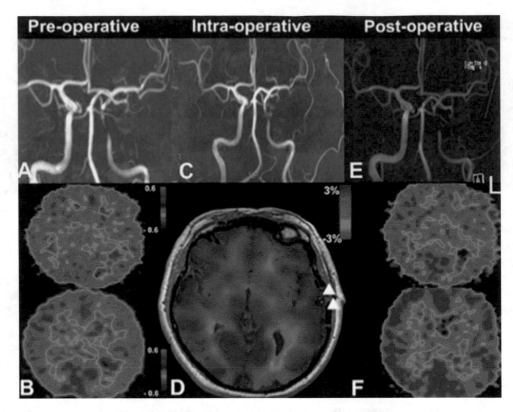

**Fig. 3** Clinical example of intraoperative BOLD-CVR assessment in a patient undergoing STA-MCA bypass for cerebral revascularization in internal carotid artery (ICA) occlusion. (**a**) Preoperative MR angiography (MRA) and (**b**) positron emission tomography (PET) in a patient with symptoms associated with hemodynamic failure due to a left intracranial ICA occlusion, who was selected for a revascularization procedure; (**c**) intraoperative MRA after STA-MCA bypass performance: according to the findings, no sufficient flow can be seen in the anastomosis, but despite this, intraoperative BOLD-CVR maps (**d**) showed high CVR values in the region adjacent to the anastomosis (white arrowheads), hinting to a good bypass function despite suboptimal MRA findings; this was confirmed by the postoperative PET (**f**) which, despite an unsatisfactory depiction of the anastomosis on MRA (**e**, red arrow), showed an improvement of the hemodynamics in comparison to preoperative findings. (Picture adapted from [24] Neurosurg Focus 2019;46:E7)

others with preserved CVR [15]. A lower BOLD signal depicts this in areas with an exhausted capacity [3]. More specifically, areas with a higher blood flow achieve a more elevated deoxyhemoglobin clearance, thereby displaying a higher oxyhemoglobin/deoxyhemoglobin ratio, which ultimately results in a higher BOLD signal [2].

Intraoperative CVR variations can either reflect the presence of residual invading tumor tissue (impaired CVR) or the correct functioning of a bypass performed for cerebral revascularization (improved CVR): BOLD-CVR could be useful in discriminating healthy brain from tumor tissue beyond the limits of current standards, like T1-weighted imaging with contrast enhancement [26], and, if future studies confirmed our preliminary results, offer a better depiction of areas of tumor infiltration. This information, intuitively, could be an adjunct tool to achieve a better extent of resection. In patients undergoing STA-MCA bypass revascularization for flow augmentation (e.g., in Moyamoya disease), immediate intraoperative feedback on the brain hemodynamic state after the micro-anastomosis could prove its efficacy in actually revascularizing the hypoperfused areas of the brain by depicting early CVR improvements, if present [24]. This information, also, could help in the operative decision-making process and estimate the success of the bypass.

Despite promising potentials, this type of information is currently not routinely available, and some limitations need further refinements. The first and most relevant issue is the obliged need for an intraoperative MR setting. Intraoperative MR assessment has not been associated with higher rates of complication, especially if standardized protocols are followed [21, 24]. However, access to this kind of facility, even though more and more common, is limited to some centers. Additionally, the technique used to produce the vasoactive stimulus (i.e., breath-holding) has inherent limitations that hinder a consistent comparison between intraoperative and pre- or postoperative data. Specifically, breath-holding allows for a rise in $CO_2$ arterial concentration, but since it also causes concomitant hypoxia, alterations of the cellular metabolism can ensue, and, therefore, $CO_2$ production might increase and influence CVR despite the examiner control [2].

Moreover, breath-holding allows for a rise in the $CO_2$ arterial concentration, which varies between subjects according to the different subject's features and physiologic parameters (age, sex, weight, metabolic status, and physical fitness) and in the same subject over time [2]. The entity of this change is unpredictable, and the vasoactive stimulus cannot be controlled or modulated [2], thereby providing different entities of vasoactive stimuli in separate investigations. This hinders a consistent inter- and intrasubject comparisons. These limitations are avoided in conscious patients by using a custom gas blender (RespirAct™, Thornhill Medical, Toronto, Canada), which allows to obtain targeted iso-oxic pseudo-square waved changes of arterial $CO_2$ concentration: a controlled and consistent stimulus can be provided by letting the subject inhale and rebreathe a gas mixture to obtain 10 mmHg raises of the $CO_2$ arterial concentration from the previously registered subject's baseline [2, 3] (Fig. 4). Nevertheless, its intraoperative employment is currently not available. With the future improvements of technologies [10, 27] used to obtain controlled changes of the $CO_2$ inhaled concentration in awake patients, we expect to be able to perform BOLD-CVR investigations with controlled vasoactive stimuli also intraoperatively in the near future (see also van Niftrik and Piccirelli [22, 23] for further technical details on the techniques used and data analysis).

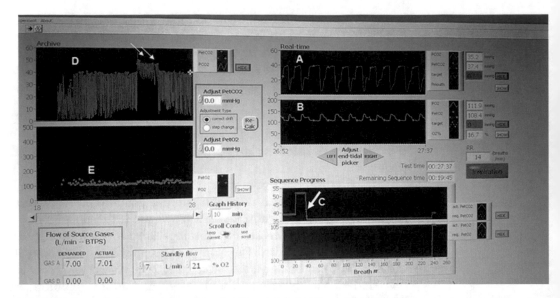

**Fig. 4** Interface of the custom gas blender (RespirAct™, Thornhill Medical, Toronto, Canada). This allows providing controlled hypercapnic stimuli of 10 mmHg $CO_2$ concentration raises and cannot be currently used intraoperatively due to hardware nuisances. Future developments of these techniques will allow us to employ it also in an intraoperative setting. (**a**) Breath-by-breath registered subject's $CO_2$ concentration and (**b**) $O_2$ concentration; (**c**) targeted $CO_2$ concentration during the exam: a 10 mmHg increase in $CO_2$ concentration of the inhaled gas is provided to the patient [arrow]; (**d**) subject's $CO_2$ concentration registered during the exam—10 mmHg $CO_2$ pseudo-square raises are obtained [double arrow], providing a valid vasoactive stimulus; (**e**) subject's $O_2$ concentration during the exam: note that the concentration remains stable during the whole exam

It is important to stress that due to the limited numbers, we are not allowed to extend our conclusions to other patients despite the consistency of our findings. To this aim, further studies with larger cohorts and extended follow-up are still needed.

Our first intraoperative BOLD-CVR assessment offered interesting results, which should deserve to be further investigated in the future: first, we observed an early intraoperative change after revascularization in patients undergoing bypass revascularization surgery for flow augmentation with BOLD-CVR. Second, we identified an association with intraoperative findings and the evolution of the disease. Third, we confirmed that this assessment is not associated with higher complication rates.

Further developments of this application will include patients with aneurysmal subarachnoid hemorrhage. Specifically, impaired CVR has been found after aneurysmal subarachnoid hemorrhage (SAH) in patients with a lower clinical grade, and progressive impairment of CVR was associated with a higher risk of delayed cerebral ischemia (DCI) [28]. The purpose of obtaining an intraoperative or ultra-early BOLD-CVR assessment would be to identify areas with impaired CVR at risk for DCI or to assess a CVR value for future comparison for follow-up, in order to predict DCI.

# References

1. Sam K, Poublanc J, Sobczyk O et al (2015) Assessing the effect of unilateral cerebral revascularisation on the vascular reactivity of the non-intervened hemisphere: a retrospective observational study. BMJ Open 5:e006014

2. Fierstra J, Sobczyk O, Battisti-Charbonney A et al (2013) Measuring cerebrovascular reactivity: what stimulus to use? J Physiol 591:5809–5821

3. Fisher JA, Venkatraghavan L, Mikulis DJ (2018) Magnetic resonance imaging-based cerebrovascular reactivity and hemodynamic reserve: a review of method optimization and data interpretation. Stroke 49:2011

4. Blair GW, Doubal FN, Thrippleton MJ, Marshall I, Wardlaw JM (2016) Magnetic resonance imaging for assessment of cerebrovascular reactivity in cerebral small vessel disease: a systematic review. J Cereb Blood Flow Metab 36:833–841

5. Mandell DM, Han JS, Poublanc J et al (2011) Quantitative measurement of cerebrovascular reactivity by blood oxygen level-dependent MR imaging in patients with intracranial stenosis: preoperative cerebrovascular reactivity predicts the effect of extracranial-intracranial bypass surgery. AJNR Am J Neuroradiol 32:721–727

6. Conklin J, Fierstra J, Crawley AP et al (2010) Impaired cerebrovascular reactivity with steal phenomenon is associated with increased diffusion in white matter of patients with Moya-moya disease. Stroke 41:1610–1616

7. Conklin J, Fierstra J, Crawley AP et al (2011) Mapping white matter diffusion and cerebrovascular reactivity in carotid occlusive disease. Neurology 77:431–438

8. Cogswell PM, Davis TL, Strother MK et al (2017) Impact of vessel wall lesions and vascular stenoses on cerebrovascular reactivity in patients with intracranial stenotic disease. J Magn Reson Imaging 46:1167–1176

9. Silvestrini M, Vernieri F, Pasqualetti P et al (2000) Impaired cerebral vasoreactivity and risk of stroke in patients with asymptomatic carotid artery stenosis. JAMA 283:2122–2127

10. Ellis MJ, Ryner LN, Sobczyk O et al (2016) Neuroimaging assessment of cerebrovascular reactivity in concussion: current concepts, methodological considerations, and review of the literature. Front Neurol 7:61

11. Urback AL, MacIntosh BJ, Goldstein BI (2017) Cerebrovascular reactivity measured by functional magnetic resonance imaging during breath-hold challenge: a systematic review. Neurosci Biobehav Rev 79:27–47

12. Mandell DM, Han JS, Poublanc J et al (2008) Mapping cerebrovascular reactivity using blood oxygen level-dependent MRI in Patients with arterial steno-occlusive disease: comparison with arterial spin labeling MRI. Stroke 39:2021–2028

13. Fraga de Abreu VH, Peck KK, Petrovich-Brennan NM, Woo KM, Holodny AI (2016) Brain tumors: the influence of tumor type and routine MR imaging characteristics at BOLD functional MR imaging in the primary motor gyrus. Radiology 281:876–883

14. Fierstra J, Conklin J, Krings T et al (2011) Impaired peri-nidal cerebrovascular reserve in seizure patients with brain arteriovenous malformations. Brain 134:100–109

15. Fierstra J, Poublanc J, Han JS et al (2010) Steal physiology is spatially associated with cortical thinning. J Neurol Neurosurg Psychiatry 81:290–293

16. Fierstra J, van Niftrik C, Piccirelli M et al (2018) Diffuse gliomas exhibit whole brain impaired cerebrovascular reactivity. Magn Reson Imaging 45:78–83

17. Fierstra J, van Niftrik C, Warnock G et al (2018) Staging hemodynamic failure with blood oxygen-level-dependent functional magnetic resonance imaging cerebrovascular reactivity: a comparison versus gold standard ((15) O-)H2O-positron emission tomography. Stroke 49:621

18. Hendrik Bas van Niftrik C, Sebok M, Muscas G et al (2019) Characterizing ipsilateral thalamic diaschisis in symptomatic cerebrovascular steno-occlusive patients. J Cereb Blood Flow Metab 40:563

19. Fierstra J, van Niftrik B, Piccirelli M et al (2016) Altered intraoperative cerebrovascular reactivity in brain areas of high-grade glioma recurrence. Magn Reson Imaging 34:803–808

20. Fierstra J, Burkhardt JK, van Niftrik CH et al (2017) Blood oxygen-level dependent functional assessment of cerebrovascular reactivity: feasibility for intraoperative 3 Tesla MRI. Magn Reson Med 77:806–813

21. Stienen MN, Fierstra J, Pangalu A, Regli L, Bozinov O (2018) The Zurich checklist for safety in the intraoperative magnetic resonance imaging suite: technical note. Oper Neurosurg (Hagerstown) 16:756

22. van Niftrik CH, Piccirelli M, Bozinov O et al (2016) Fine tuning breath-hold-based

cerebrovascular reactivity analysis models. Brain Behav 6:e00426

23. van Niftrik CHB, Piccirelli M, Bozinov O et al (2017) Iterative analysis of cerebrovascular reactivity dynamic response by temporal decomposition. Brain Behav 7:e00705

24. Muscas G, Bas van Niftrik CH, Fierstra J et al (2019) Feasibility and safety of intraoperative BOLD functional MRI cerebrovascular reactivity to evaluate extracranial-to-intracranial bypass efficacy. Neurosurg Focus 46:E7

25. Davis TL, Kwong KK, Weisskoff RM, Rosen BR (1998) Calibrated functional MRI: mapping the dynamics of oxidative metabolism. Proc Natl Acad Sci U S A 95:1834–1839

26. Hsu YY, Chang CN, Jung SM et al (2004) Blood oxygenation level-dependent MRI of cerebral gliomas during breath holding. J Magn Reson Imaging 19:160–167

27. Fisher JA (2016) The CO2 stimulus for cerebrovascular reactivity: fixing inspired concentrations vs. targeting end-tidal partial pressures. J Cereb Blood Flow Metab 36:1004–1011

28. Carrera E, Kurtz P, Badjatia N et al (2010) Cerebrovascular carbon dioxide reactivity and delayed cerebral ischemia after subarachnoid hemorrhage. Arch Neurol 67:434–439

# Correction to: Recent Advances and Future Directions: Clinical Applications of Intraoperative BOLD-MRI CVR

## Giovanni Muscas, Christiaan Hendrik Bas van Niftrik, Martina Sebök, Alessandro Della Puppa, Luca Regli, and Jorn Fierstra

Correction to:
Chapter 11 in: Jean Chen and Jorn Fierstra (eds.), *Cerebrovascular Reactivity: Methodological Advances and Clinical Applications*, Neuromethods, vol. 175, https://doi.org/10.1007/978-1-0716-1763-2_11

The authors name Alessandro Della Puppa is wrongly listed as "Puppa A.D" and this has been updated.

---

The updated online version of this chapter can be found at
https://doi.org/10.1007/978-1-0716-1763-2_11

Jean Chen and Jorn Fierstra (eds.), *Cerebrovascular Reactivity: Methodological Advances and Clinical Applications*, Neuromethods, vol. 175, https://doi.org/10.1007/978-1-0716-1763-2_12, © Springer Science+Business Media, LLC, part of Springer Nature 2022

# INDEX

**B**

BOLD fMRI.................................... 76, 81, 83, 90–92, 95,
   97, 98, 104, 112, 134, 145, 150, 160, 167–170,
   172, 181
Brain aneurysm.......................................................... 44
Brain gliomas.................................................... 167–181
Breath-holding .............. 20–22, 24–26, 28, 64, 66, 124,
   167–181, 188, 209, 213

**C**

Carbon dioxide (CO$_2$)............................6, 60, 63, 90, 94,
   124, 136, 170, 178
Cerebral blood flow (CBF) .................. 1–10, 12, 19–23,
   25–27, 33–37, 39–46, 48, 51–53, 60–67, 77, 78,
   80, 82–84, 90, 92, 93, 95–98, 106, 107,
   109–112, 124, 127, 129, 132, 134, 167, 170,
   172, 185–187, 191–193, 197–201, 207, 208, 211
Cerebral ischemia ......................... 96, 157, 169, 186, 214
Cerebrovascular disease (CVD)................... 6, 21, 23, 44,
   45, 48, 49, 52, 119–161, 185–187, 189, 191,
   193, 195, 201
Cerebrovascular reactivity (CVR) ................ 1–16, 19–29,
   33–54, 59–70, 76–85, 90–99, 103–113, 124,
   167–181, 185–201, 207–215
Cerebrovascular reserve .................... 23, 28, 68, 96, 105,
   130, 186, 198–200, 207, 211

**E**

Echo planar imaging .................... 91, 136, 140–142, 146

**F**

Functional MRI (fMRI) ................75–85, 90–92, 95, 97,
   98, 104, 109, 112–113, 119, 120, 132–134, 136,
   141, 143–148, 150–152, 157, 158, 160, 161,
   167–174, 177–179, 181

**G**

Gliomas................................................................ 167–181

**H**

Hemodynamic failure .........................50, 92, 96, 97, 212
Hemodynamic reserve (HR) .............................. 119–161

**H**

Hemodynamics ............................... 9, 24, 28, 45, 60, 66,
   67, 82, 90, 93, 95, 96, 99, 106, 121, 129–133,
   168, 169, 172, 178, 188, 193, 196–198, 200,
   207–209, 211–213
HMPAO ................................................................48, 50
[$^{15}$O]H$_2$O .....................................................34, 36, 38–53

**I**

Intracranial arteriosclerosis ........................................ 44
Intra-operative ..................................................... 208–214

**K**

Kinetic modelling................................................41, 43, 51

**M**

Magnetic resonance angiography
   (MRA)....................................120–127, 132, 140,
   146, 147, 156–159, 212
Magnetic resonance imaging
   (MRI)................................... 20, 21, 23, 24, 26, 27,
   37, 38, 42, 46–53, 64, 65, 77, 90–92, 106–108,
   111, 133, 141, 142, 146, 159, 167, 170, 171,
   188, 189, 192–196, 198, 200, 201
Moyamoya disease (MMD)............................11, 44, 46,
   47, 82, 120, 131, 132, 146, 147, 150, 152–154,
   156, 157, 159, 160, 196, 210, 213
MR perfusion ....................................................... 136, 141

**N**

Neuroimaging ............................... 81, 185–187, 191–193
Neurovascular uncoupling (NVU) ..................... 167–181

**P**

Perfusion reserve ................................................. 90, 96–99
Positron emission tomography (PET)............. 34–53, 65,
   96, 108, 134, 191, 192, 201, 212

**R**

Rebreathing .............. 20, 22, 26, 28, 143, 190, 191, 195
Resting state fMRI (rs-fMRI) ............... 80–85, 178, 179

## S

Single photon emission computed tomography
(SPECT) ...............................34, 48–51, 108, 119,
190, 192, 196, 198, 201
STA-MCA......... 132, 134, 157–160, 198, 209, 211–213
Strokes ................. 3, 4, 6, 9, 46, 82, 85, 90–97, 99, 110,
119, 120, 130, 131, 133, 134, 147, 157, 161,
185, 192, 195–198, 207
Subarachnoid hemorrhage (SAH) ......... 61, 67, 169, 214

## T

Transcranial doppler (TCD).......................59–69, 89–99,
108, 192, 196, 198, 199, 201
Traumatic brain injury (TBI) .................. 50, 65–68, 195

## U

Ultrasound........................................ 61, 63, 91, 111, 192

Printed in the United States
by Baker & Taylor Publisher Services